Are You Smarter Than A Chimpanzee?

Test yourself against the amazing minds of animals

BEN AMBRIDGE

P

PROFILE BOOKS

For my two favourite animals

First published in Great Britain in 2017 by
PROFILE BOOKS LTD
3 Holford Yard
Bevin Way
London WC1X 9HD
www.profilebooks.com

1 3 5 7 9 10 8 6 4 2

Typeset in 11.25/13.75 Chaparral Pro
Designed by Nicky Barneby @ Barneby Ltd
Printed and bound in Great Britain by Clays, Bungay, Suffolk

The moral right of the author has been asserted.

A CIP catalogue record for this book is available from the British Library.

ISBN 978 1 78125 573 5
eISBN 978 1 78283 216 4

FSC
Mixed Sources
Product group from well-managed
forests and other controlled sources

Cert no. SGS-COC-2061
www.fsc.org
© 1996 Forest Stewardship Council

The paper this book is printed on is certified by the © 1996
Forest Stewardship Council A.C. (FSC). It is ancient-forest friendly.
The printer holds FSC chain of custody SGS-COC-2061

Contents

Introduction: Relative Values   1

An Expensive Cappuccino   5

The Eye of the Beeholder   9

The Usual Waspects   12

Spot the Difference   15

How the Giraffe Got His Neck   18

A Fishy Tale   20

Fish Are Amazing   21

A Fishy Tale: You Found the Word   25

A Fishy Tale: You Didn't Find the Word   26

Are You Big-Headed?   29

Goatbusters   32

Pigeon's Treat   36

MMMMonkeying Around   43

More Monkeyconomics   47

Ape, Man United   51

Having a Mare (or not)   56

A Stable Personality?   58

The Truth about Cats and Dogs   65

Dog-Person or Person-Dog?   67

A Walking Dogtionary   70

A Walking Dogtionary: Test   72

A Walking Dogtionary: Two Years Later   75

Who Man Being?   76

A Formidable Problem   82

The Tower of Han(t)oi   86

Losing Your Marbles   91

Counting Cheep   94

Circle of Life   97
Pige-lusi-on #1   100
Pige-lusi-on #2   103
Mathemalex   106
Parrotnormal Activity   108
A Shaggy Dog Story   112
Dogtanian and the Three Must-Get-Pairs   116
Turning Japanese   121
Why Can't We Talk to the Animals?   124
Something To Crow About?   127
Students vs Squirrels, Sorta   132
Box Clever   135
Is There a Dog-tor in the House?   138
Let's Get Rat-Arsed   142
Oh, What a Tangled Web We Weave   145
Insex   150
Let's Spend Some Quaility Time Together   154
A Sick Idea   157
You've Got To Hide Your Bug Away   160
Should I Stay or Should I Go?   163
Great Tits and Seedy Locations   167
Great Tits and Seedy Locations: Test   168
The Pecking Order   171
A Weighty Problem   175
For Eagle-Eyed Readers   177
Fowl Play   179
For Eagle-Eyed Readers: Test   182
Are You a Bat, Man?   183
An Elephant Never Forgets   187
A Shrewd Judgement   189
Left Behind?   193
Cuckoo Clocks His Rivals   198
Dad Calls It Quits   203

School of Collective Nouns 206

Signed, Sea-Lioned, Delivered 209

Bird Is the Word (or Starlings in Their Eyes) 211

Stringing You Along 216

Are You Smarter than an Orang-Utan? 219

Are You Smarter than an Orang-Utan? Test 220

Are You Smarter than a Chimpanzee? #1 222

Are You Smarter than a Chimpanzee? #2 225

Are You Smarter than a Chimpanzee? #3 228

It's Been Emotional 231

Psycho Gorilla, Qu'est-ce que c'est? 237

Every Body Hurts? 242

Oh, Beehave 246

Taking the P*** 249

Self-Awareness: Have Monkeys Cottoned On? 251

Epilogue: Relative Values Revisited 254

References 262

Credits 280

Introduction: Relative Values

Are you smarter than a chimpanzee? Cleverer than a cat? Brainier than a bat? More perspicacious than a pigeon?

Well of course you are! After all, you are a *Homo sapiens* ('wise man'), while the cats, bats, rats, hogs, dogs, frogs and all of the rest of it are 'just animals', right?

The idea that human beings are somehow different from other animals has a long history. Take, for example, the Bible. It's right there on Page 1:

And God said, Let us make man in our image, after our likeness: and let them have dominion over the fish of the sea, and over the fowl of the air, and over the cattle, and over all the earth, and over every creeping thing that creepeth upon the earth. (Genesis 1:26)

For most of our history, religious scholars have argued that, since man alone is made in God's own image, our species is unique in having thoughts, consciousness, even an immortal soul. In our secular modern world, this idea of *human exceptionalism* is no longer the sole preserve of the religious. As bioethicist Wesley J. Smith asks:

What other species has been able to (at least partially) control nature instead of being controlled by it? What other species builds civilizations, records history, creates art, makes music, thinks abstractly, communicates in language, envisions and fabricates machinery, improves life through science and engineering, or explores the deeper truths found in philosophy and religion?

When you put it like that, human exceptionalism starts to sound like simple common sense. But my goal in this book – by having you compare yourself against our animal cousins on a range of psychological tests – is to encourage you to say, 'Hang on a minute! Are we *really* so different?'

While we may not choose to call them 'civilisations', many animals – from chimpanzees to chickens – live in groups with a clearly defined pecking order (in which we'll find your own position), and even ants and bees even get to vote (and, as a result of this democratic swarm-intelligence, will give you a good run for your money on tests of route-planning and puzzle-solving). Starlings 'make music' in that their songs are constructed around the same scales as most traditional Western compositions (which allows you to compete against one in a sing-off). 'Abstract thinking' is shown by crows, squirrels and box turtles (and *you*?) in tests that involve using patterns or rational inferences to figure out the location of a tasty treat. Whether or not other animals can learn language is a long-running debate (which we'll explore by trying to teach you Japanese), but many – dogs in particular – can learn an impressive number of words (a feat you'll attempt for Russian). And while it might be a stretch to call it 'science and engineering', chimpanzees certainly use tools, for example when *ant-dipping* (using a shoot as a spoon to pick up ants) and *termite-fishing* (using a thin twig as a rod to catch termites) – phenomena that hold the key to understanding human handedness (on which you'll compare yourself to your cat, as well as a northern tree shrew).

More generally, we'll see that almost all of our human abilities, activities and concerns – from choosing a good-looking partner and understanding their facial expressions (at which you'll compete against bees and chimpanzees ...) to getting a bargain and learning to quit while you're ahead (... pigeons, guinea pigs and hairy armadillos) – boil down to one of what biologists call 'the four Fs': fighting, fleeing, feeding and – ahem – fornicating.

Of course, nobody is denying that humans – even you – can do plenty of things that other animals can't (so don't worry, you're not going to lose on *all* the tests). All I hope to persuade you is that, in the words of Charles Darwin (1809–1882), the difference is 'one of degree and not of kind': the same abilities that allow starlings to sing, parrots to count and fish to find their way home allow humans to write symphonies, do calculus and invent Google maps. We don't do anything *different* from other animals; we do the same things, only better.

While some of these tests might sound a little frivolous – and I certainly hope you'll find them a lot of fun – all have a firm scientific basis, and are based on peer-reviewed articles in reputable academic journals. But just *why* are respectable scientists in the business of comparing humans and animals on everything from susceptibility to visual illusions (baboons and baby chicks) to the likelihood of winning the star prize in a game-show finale (pigeons)?

The answer is that, by exploring the similarities and differences between humans and other animals, we can begin to understand when and how our abilities, our likes and dislikes, and even our foibles and mental blind spots arose in the course of evolution. It was this comparative method, most famously applied to finches, that led to Charles Darwin's theory of evolution by natural selection in the first place. Contrary to popular belief, the idea of evolution wasn't invented by Darwin. At least, not by *Charles* Darwin. His grandfather Erasmus asked in *Zoonomia* (1794):

From thus meditating on the great similarity of the structure of the warm-blooded animals, and at the same time of the great changes they undergo both before and after their nativity; and by considering in how minute a portion of time many of the changes of animals [. . .] have been produced; would it be too bold to imagine, that in the great length of time, since the earth began to exist, perhaps millions of ages before the commencement of the history of mankind, would it be too bold to imagine, that all warm-blooded animals have arisen from one living filament?

What Charles Darwin did invent, or at least popularise,* was the idea of evolution *by natural selection*: that evolution is the result

---

* In fact, as Darwin later acknowledges, the idea was first mooted by Scottish fruit farmer Patrick Matthew, who buried it in an appendix to his 1831 smash hit *On Naval Timber and Arboriculture*.

of animals with favourable traits successfully reproducing, while those with less favourable traits die out. Before *On the Origin of Species*, the dominant idea – as discussed in *Zoonomia*, as well as in Jean-Baptiste Lamarck's *Philosophie Zoologique* (1809) and Robert Chambers's *Vestiges of the Natural History of Creation* (1844) – was 'use it or lose it': the body parts and mental abilities that animals use frequently become strengthened – and those they don't, weakened – with these changes passed on to subsequent generations. Indeed, this idea was not completely abandoned until natural selection was wedded to Mendelian genetics (the so-called *modern synthesis*) at the start of the twentieth century; *Origin* was merely the first nail in the coffin.

As you will soon see for yourself as you take these tests, and learn about the science behind them, twenty-first-century findings in evolutionary biology, psychology and genetics have confirmed the modern synthesis beyond reasonable doubt. Darwin was right: when it comes to the differences between humans and other animals, everything is relative and everything is *a* relative: we are all part of one big family.

An Expensive Cappuccino

What better place could there be to snuggle up with this book and begin your journey through the animal kingdom than in a cosy coffee house? But, oh no, you're in an unfamiliar town, which has somehow managed to resist the onslaught of the big chains. You're going to have to take your chances with an independent. And, what luck, here are three, right on the same street! They are all pretty interchangeable in terms of their décor and ambience, so all you have to go on is the price of your cappuccino:

£0.80 in Coffee House A
£1.80 in Coffee House B
£2.80 in Coffee House C

Which do you pick?

ANSWER

Have you ever seen a capuchin monkey? They live in South and Central America, and are brown all over, except for a light chest and face, topped off with a little brown hood. This distinctive appearance gives the monkey its name. Capuchins are named after an order of friars (*Ordo Fratrum Minorum Capuccinorum*) who wear distinctive brown robes with large hoods. So, when a type of coffee with almost exactly the same shade of brown came along, it was perhaps inevitable that it would be called a *cappuccino*, or 'little Capuchin' (after the monk, not the monkey).

All of which brings us nicely back to your cappuccino. Which one did you pick? Probably not the cheapest, right? In fact, when placed in scenarios of this type, most people – provided they can afford it – go for the most expensive option. Why? In the absence of any other information to go on, price is generally a fairly reliable indicator of quality.

Indeed, when a product is discounted – even for completely innocuous reasons – we just can't shake the feeling that it is somehow worse. One study found that participants who had paid $1.89 for an energy drink showed a greater mental boost on a set of puzzles than participants who had paid just $0.89 for the drink, even though they were given the brand name (*SoBe Adrenaline Rush*) and told that the discount was possible because the drinks had been bought using the university's institutional discount (and not, for example, because the drink is a below-par product that the manufacturer is trying to offload). The researchers also found that participants rated the same 'painkiller' (actually a placebo pill) as more effective if it had been purchased (in this case, by the researchers) at full price ($2.50) rather than at a hefty discount ($0.10).

Where do the capuchins come in? You guessed it: a group of researchers decided to investigate whether our primate relatives also show these pricing effects. First, the monkeys were given the opportunity to sample the goods on offer – differently flavoured orange and blue ice chunks – with any who showed a strong preference for one or the other removed from the study. Next, the monkeys were taught that orange chunks were 'cheap' and blue chunks 'expensive' (or vice versa for half of the monkeys): each time the monkey handed the experimenter a token, he received in exchange either three of the cheap chunks or one of the expensive chunks (determined by the experimenter). Finally, the monkeys were given free rein at an ice-chunk buffet. If they had come to associate price with quality, they should have gobbled up the fancy expensive ice, while turning their noses up at the cheap crap. Indeed, this is exactly what humans do if placed in a similar scenario (though with – usually – wine, rather than ice chunks).

The capuchins, on the other hand, showed no preference whatsoever for the more expensive product. In fact, whatever the experimenters tried – jelly shapes, identical cereal with different branding, an enforced waiting period for certain foods – the monkeys stubbornly refused to prefer the expensive goods. And it's not simply that they failed to learn about the different prices: if, instead of a free buffet, the monkeys were required to pay for their treats, they overwhelmingly chose the cheaper option, in order to maximise their limited budgets. Capuchins are perfectly capable of learning that one product is cheaper than another; they just don't consider the more expensive one to be in any way better.

That's all very well, but what – you may be asking – was the point of running this study with capuchins? Were the experimenters just having a laugh? Not at all. The researchers were following the method of *comparative psychology*. When we want to understand why humans show some particular phenomenon (in this case, associating price with quality, even when it makes little sense to do so), a useful approach is to compare humans with a species that is similar in relevant ways. In this case, capuchins fit the bill, because they are capable of learning about prices and using them to guide purchasing decisions. If, despite these similarities, the other species does not show the same pattern, we can conclude that the cause of the phenomenon in humans is probably something that is unique to our species.

In this case, the best candidate for that something is experience with markets. We humans have learned that, due to the laws of supply and demand, if someone can get away with charging more for something, it just *must* be a better product than its rivals, even if we find it hard to tell the difference. (Just think of Apple, which became the world's most successful company, despite – or perhaps even *because of* – its high prices.) Having formed an expectation that high price equals high quality, we can't help applying it to situations in which we know it is not relevant (e.g., a university researcher buying an energy drink using an institutional discount).

So, if you're on a limited budget but find it hard to resist the premium brands when it comes to your morning cappuccino, perhaps you should delegate the coffee run . . .

. . . to a capuchin.

The Eye of the Beeholder

So, it's Monkey 1–0 Man in our battle of the species. But there's no shame in that; after all, capuchin monkeys are pretty close to us in evolutionary terms. So, now let's see how much you have in common with one of your lowliest, much more distant, relatives: the humble bumblebee. Below are six pairs of pictures. Your task is simple: for each pair, pick the one you prefer, based on your immediate gut reaction.

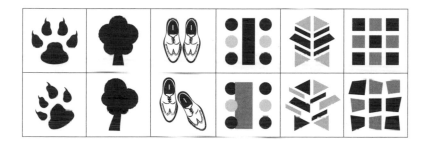

ANSWER

As you may have begun to guess as you worked your way through the pairs, this test is about symmetry. Overall, people generally show a small but reliable preference for symmetrical over asymmetrical figures – a preference that holds for both real objects (e.g., the three pairs on the left) and imaginary, abstract objects (the three pairs on the right).

But how did this come about? Did we 'learn' to prefer symmetrical objects, or were we born with this preference already built in? In principle, we could find out by raising a group of children, without ever letting them see anything symmetrical, and looking to see if they also showed this symmetry preference. In practice, of course, this would be impossible. Even if the study somehow got past a university ethics committee (which it wouldn't, as it would involve preventing babies from seeing any human faces), these poor children would be so unusual – in terms of both their visual and their psychological development – that they could tell us almost nothing about the origins of a symmetry preference in typical development.

When a human study is impossible, an animal study is usually the next best thing. It is neither particularly difficult nor particularly cruel to keep, for example, baby bumblebees away from symmetrical patterns until they are ready to start foraging. When they do so, it is a simple matter to offer them the choice of symmetrical or asymmetrical flowers (or, more usually, computer-generated patterns) and see which they pick. When this is done, it turns out that, just like adult humans (and also many bird and fish species), bees prefer symmetrical patterns, which suggests that the preference is hard-wired, rather than learned.

But why do we (and many other animals) prefer symmetrical patterns? One possibility is that they are somehow easier to process. If you look back at the pairs on the previous page, you'll find that, in every case, you can get an idea of what's going on (visually speaking) just by glancing at the top picture in each pair, while the bottom picture demands more deliberate inspection. Because the visual system, like most brain systems, is inherently 'lazy', we prefer symmetrical patterns because – so this theory goes – they require less work.

A second possibility is that our preference for symmetry is a consequence of evolution. For things that *should* be symmetrical, such as human faces and many pollen-producing flowers, any deviation from symmetry often indicates poor-quality genes and – as a result – a poor-quality individual. In humans, symmetry is associated with intelligence, athleticism and resistance to depression (and, in flowers, with greater pollen production). So, according to this theory, we have evolved a preference for symmetry because this maximises our chances of picking a mate with good genes, and therefore successfully reproducing; our preference for symmetrical abstract patterns (and a symmetrical drawing of a pair of shoes) is an evolutionary hangover from our preference for symmetrical faces.

So, which theory is correct: ease of processing or evolutionary preference? The jury is still out, but one piece of evidence from the original version of the study you just completed seems to favour the latter. When the researchers broke down the results by gender, they found that only men showed a reliable preference for the symmetrical over asymmetrical figures (63 per cent to 37 per cent), with women's choices almost exactly 50/50. This is bad news for the ease-of-processing theory, as it seems extremely unlikely that male and female brains differ in something as basic as visual processing (or, to give it its non-technical term, 'seeing'). Now, this gender difference isn't knock-down evidence for the evolutionary-preference theory either (in face-rating studies, both men and women show a symmetry preference), but it seems at least consistent with it: we know that, when it comes to choosing a mate, men of all cultures place greater importance on physical attractiveness than do women (a finding I explored in my last book, *Psy-Q*), so it's possible that only for men does one characteristic of physical attractiveness – facial symmetry – spill over into inanimate objects.

The Usual Waspects

Sticking with the subject of faces, let's move on from the hon-ey-producing heroes of the insect world to the villainous *vespidae*. Many people say that, while they may be terrible with names, they never forget a face. But is that still true when the face in question belongs to a wasp?

Below is the face of a criminal wasp who was caught – ahem – in a police sting operation. Study his face carefully as, on the next page, you will be asked to pick him out of a line-up.

ANSWER

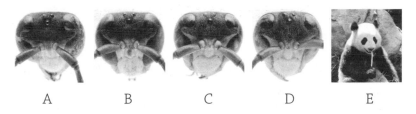

A	B	C	D	E

Did you spot him? I hope you didn't pick out Gao Gao the panda, who was just a stooge in this identity parade (but who we'll come back to later in the book). The answer is at the bottom of the page. If you need a hint, the key distinguishing feature in this line-up is the amount of black pigmentation in the area around the ocelli (eyes), ranging from least (Wasp A) to most (Wasp D).

Many very simple species, even cockroaches and amoebas, can distinguish kin (i.e., close genetic relatives) from non-kin. However, as late as the 1960s, it was generally thought that only vertebrates (species with a backbone) had the brainpower necessary to recognise individuals that they have met before. More recently, there has been some evidence to suggest that certain species of beetle, crayfish and hermit crab may also have this ability, although the issue remains controversial.

Wasps, on the other hand, are the individual-recognition champions of the invertebrate world. In addition to an odour shared by all members of the colony, many species of wasp can recognise familiar individuals by their faces. In fact, a recent study, which involved presenting nest-mates and outsiders to a colony, found that, if they can see the face, wasps ignore the odour cue altogether. That is, if an unfamiliar face approaches the nest, it is attacked, even if it has been artificially scented with the odour of that colony.

Now, if this isn't too personal a question, which of the wasps above do you find most attractive (they're all males, by the way)? In a scientific paper with an unusually racy title – *Sexy Faces in A*

Wasp C is the guilty party

Male Paper Wasp – a group of Brazilian scientists found that female wasps went crazy for males with large black spots (D), but stung, bit or flew away from those who did not (A). In fact, the scientists found that they could easily improve a wasp's prospects with a dab of black paint (or just as easily destroy them with a dab of brown). This was big news because, while sexually selected traits are common in many animals – just think of the peacock's plumage – they have rarely been found in insects.

Speaking of sexy peacocks . . .

Spot the Difference

Which of these peacocks do you think enjoys greater success with the ladies?

ANSWER

If you said the top one, on the basis that it has more eyespots, then you're right. Probably.

Whether or not peahens (that's the females) prefer peacocks (males) with more eyespots has proved a surprisingly difficult question to answer. A 2011 analysis that pooled the data from all previous studies found that it's not a simple case of the more (eyespots) the merrier (the male peacock). What is important is just having *enough*. Females almost never mate with males that have fewer than 144 eyespots; but provided a male meets that magic number, there's no benefit in having more. It is true that peacocks unfortunate enough to have a number of their eyespots removed by a sneaky experimenter suffer badly in the mating game; but only because the experimenters usually snip off around twenty-five spots, enough to leave the peacock – like our poor friend on the previous page – just shy of the magic number of 144. (Males typically grow 169 eyespot feathers, but have invariably lost at least a few to natural causes before the experimenter turns up with the scissors.)

But why is there a magic number at all? The reason is not, of course, that peahens are actually counting the number of feathers. Rather, having so many feathers removed – far more than would usually fall off naturally in the wild – just makes a peacock look, to use the technical term, a bit odd. The removal of twenty-five feathers affects the size and symmetry of the train – both things that peahens seem to care about a lot.

While it's unclear whether or not peahens are bothered about the quantity – per se – of a peacock's eyespots, they are certainly bothered about the quality. Most important to mating success are the hue (roughly speaking, colour) and iridescence of the blue-green patch (an iridescent surface is one that appears to change colour as it moves).

Why do peahens care about eyespots at all? The answer, as Darwin had figured out by his 1871 follow-up to *On the Origin of Species* (1859), is sexual selection:

We may conclude that the pairing of birds is not left to chance; but that those males, which are best able by their various charms to please or excite the female, are under ordinary circumstances accepted. If this be admitted, there is not much difficulty in understanding how male birds have gradually acquired their ornamental characters. All animals present individual differences, and as man can modify his domesticated birds by selecting the individuals which appear to him the most beautiful, so the habitual or even occasional preference by the female of the more attractive males would almost certainly lead to their modification; and such modifications might in the course of time be augmented to almost any extent, compatible with the existence of the species.

In other words, if just a few females have just a slight preference for males with fancy eyespots, the genes of those males – including the genes that produce fancy eyespots – will be more likely to be passed on than the genes of their less fortunate competitors.

Darwin's version of evolution is so widely accepted today* that it is difficult to believe that it was neither the first nor, on the surface, the most plausible . . .

---

* That said, not all biologists accept the sexual-selection aspect of Darwin's theory; see, for example, http://science.sciencemag.org/content/311/5763/965.full

How the Giraffe Got His Neck

Rudyard Kipling never wrote a *Just So Story* called *How the Giraffe Got His Neck*, but before Charles Darwin came along, the leading theory of evolution was one that could have been lifted straight off its pages: giraffes stretch their necks to reach high-up leaves, and this exercise causes their necks to become stronger and longer. This characteristic is then passed on to their offspring.

Although, as we saw in the Introduction, this idea of *soft inheritance* was mooted in both Erasmus Darwin's *Zoonomia* (1794) and Robert Chambers's *Vestiges of the Natural History of Creation* (1844), it has subsequently become most strongly associated with the French biologist Jean-Baptiste Lamarck (1744–1829); indeed, it is often referred to as *Lamarckism*. Although, with the benefit of hindsight, this theory seems näive in the extreme, it is not difficult to see how, on the face of it, it seems much more plausible than Charles Darwin's alternative: essentially that (a) some giraffes *just so happen* to have necks that are longer than others, and (b) these giraffes are more likely to mate successfully; neither of which seems particularly likely. Indeed, the superficial implausibility of Darwinian evolution is probably one of the main reasons why a vocal minority refuse to believe it today, even though modern genetic science allows us to map out directly the associated changes in the genome, including – in 2016 – those responsible for the giraffe's long neck.

But just why does a long neck increase a giraffe's chance of successfully mating (and so passing on its long-neck genes to its offspring)? The textbook story is that giraffes with longer necks are able to reach more leaves than their rivals, and so are less likely to starve before a mating opportunity comes along. However, this isn't necessarily true. For one thing, when competition for food is at its fiercest (during the dry season), giraffes feed mainly from low-lying shrubs. Indeed, giraffes usually feed with their necks bent rather than fully extended. And if it's all about reaching tall trees, why have neck lengths increased faster than leg lengths? Longer legs would be a 'cheaper' way of achieving more height, as pumping blood to the brain up a long neck is difficult and costly in terms of energy.

These types of argument led to a rival hypothesis: sexual selection. Males fight for females by clubbing each other with their heads, and a longer neck means more clubbing power. This idea has gained quite a bit of traction, to the extent that some people think that the 'reaching for leaves' hypothesis has been thoroughly debunked. However, the sexual selection hypothesis was itself debunked by a study which found that males don't have longer necks than females, which they would if long necks had emerged to help males fight over mates. Support for the 'reaching for leaves' idea also comes from a study which found that, while giraffes might not *often* stretch for the highest leaves, they achieve the most biomass per bite when they do, perhaps because these leaves are the most nutritious.

Meanwhile, Lamarckian evolution has been enjoying a resurgence, of sorts, in the form of the emerging field of *epigenetics*. It turns out that, although the environment can't change our genes or DNA per se, it can change chemical tags that turn these genes on or off. Are these changes heritable? In plants and worms, yes. In mammals, probably not: most of these tags are reset during two separate rounds of sex-cell 're-programming'. Not that this message has got through to much of the media (the fact is, sex cells). So if you read a newspaper story about how susceptibility to obesity or heart disease in humans can be acquired from the environment and passed on through epigenetic tags, it is almost certainly – like the Lamarckian story of how the giraffe got his long neck – nothing more than a tall tale.

A Fishy Tale

Now, from a tall tale to a fishy tale. Below is a wordsearch. Your job is to find a mystery eight-letter word that relates, in some way, to the topic of animal psychology. Your ability, or inability, to find this word tells us something very important about your psychology.

```
TKAHPDASKKALAALOLDZRVTVATGUPPENTO
CUIPDFNOSYLZORDFBFLYPYLPIYKDMGUJNY
PFVBICWJGKIKBAJDDALMAGYHCABCNKTAV
VFGAJNZVFRBJCYVWTPSFPDTFVCDMODICU
OOAPALJOKBPVKMNRVADJGLKHAJTBASOSU
FELHCSIJEFDTHNPAPNPRGHOLARITYNFORR
OCOSCGWYSASLOUIWWYPPAAAGHBMYAREN
YSIOOPNMNDTVCANZVOSKRBEDKNKVUPAP
ZKETZASFDCJYATAPNVNWHIAFHDTMOKAZZ
KPPRTINOAPVLLYSOMNTJNKUEZONIARENA
HDGTAIYPMOKUHVVYZNONYOPWHMNDYIYL
WENYOPZOTCOURHOJMNIAOYMKMHUMLOA
HSKLWAOULAGHOYOVDBSUKPNEZSOIHONP
GKTOYFNOKZARKPEYZNOTOCNGACRYOVLO
ZSOGODSOTOWOODZFSYYDOEPSRCTGEIDNI
YOMMVWPWYJYNMTMMPSBREFNEOPZBKIM
YYCWHLAHFLDIOUPAOKGTUZNYUKRJNWSO
PVEISVLOYTWSCGWREAAGKBBNPECLJPOZC
SPCLEBECMPGOYHSNHOMZJAHHASSKBHBG
AITUFDOSNGWANCJHGNSMLACKHOSMJPVZ
```

If you found the word, congratulations! Now turn to page 25.

If you didn't manage to find the word, hard luck. Now turn to page 26.

Fish Are Amazing

Another fishy tale now, and, with it, more evidence for the Darwinian version of evolution. This evidence relies on the fact that many species of fish can learn a simple route. But can you?

The map below shows a route through Liverpool city centre to the docks. Study the map for a minute or two, then turn the page and use a pencil to recreate the route exactly on the blank map.

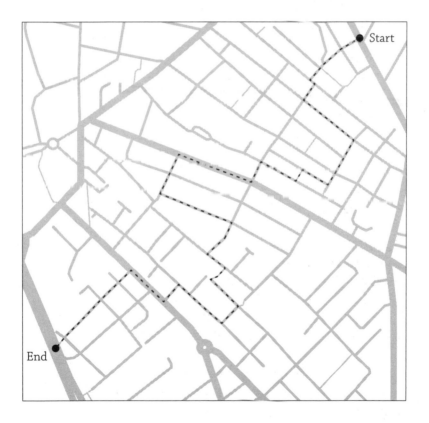

ANSWER

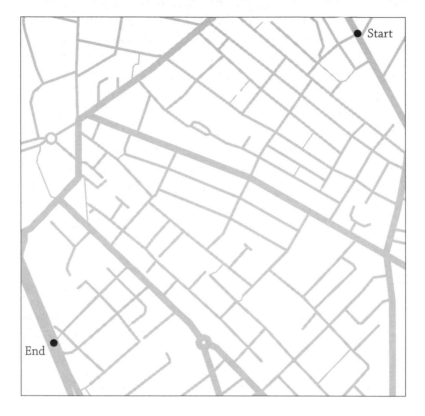

After checking your performance, you might like to compare yourself to a few of your friends and family. Interestingly, studies of this type have consistently found that males outperform females on this task (and also rely less on the use of landmarks). Although the difference in performance is small, it is comfortably statistically significant (the concept of statistical significance is explained in the section Parrotnormal Activity). The same gender difference was found when a group of Scottish biologists gave a similar – though, of course, much simpler – spatial learning task to rose bitterling, a species of freshwater fish found in Japan and China.

While this is an interesting finding, the researchers were not actually looking for gender differences per se. In fact, the study was designed to investigate a much more interesting and important question: is intelligence heritable? That is, if you are bright, are your kids likely to be bright too? Although plenty of studies have found evidence for heritability of IQ in humans (see Stuart Ritchie's *Intelligence: All That Matters*, for a very readable review), many people object that it is difficult to factor out non-genetic differences completely: people with high IQ are more likely to provide a stimulating intellectual environment for their children, in the form of books, conversation, good schools, private tutors etc. Rose bitterling do not read to their children or worry about school catchment areas, and so make excellent test subjects for researchers interested in the heritability of intelligence. Furthermore, because they're fish, we can control who has kids with whom in ways that would be wholly unfeasible – not to mention unethical – with humans. That is, if intelligence really is heritable, we ought to be able to create 'bright' or 'dim' children by breeding two 'bright' or 'dim' adults respectively.

So can we? Yes! The offspring of adults who did well on the spatial learning task (finding food in a simple maze) performed better at the same task than did the offspring of adults who did poorly. Furthermore, thanks to some clever DNA analysis, the researchers were able to show that the intelligence of both the mummy and the daddy fish made an independent contribution to the brainpower of little Nemo.

How is intelligence passed on in the wild? Simple: sexual selection, just like we saw for the wasp's black spots and the peacock's fancy plumage. Fish that do better at this spatial learning task are more likely to successfully reproduce than those that do poorly. But why? Not, of course, because fish 'prefer' more intelligent partners (what are they going to do – give them an IQ test?). For rose bitterling, the answer lies with this species' rather unusual mating habits, which revolve around mussels (not muscles; mussels, as in the food). The female places her eggs inside the mussel (via the mussel's 'out' tube), and the male ejaculates over the end of the mussel's 'in' tube, which ensures that the sperm are sucked in. Males adopt two different tactics. Large, dominant males ('guarders') jealously guard both females and mussels, and don't ejaculate until the former has deposited her eggs in the latter. Smaller, less dominant males ('sneakers') look for mussels in which they think a female is about to deposit her eggs, and get their sperm in there first (sperm can survive for about a quarter of an hour inside a mussel). The 'guarder' and 'sneaker' roles aren't fixed: a given male will adopt a guarding strategy if his nearby rivals are smaller than him, and a sneaking strategy if they are bigger.

So, the reason that rose bitterling with greater spatial intelligence (as measured by the maze test) are more likely to reproduce successfully is that they are better at learning and remembering where the most promising mussels are to be found. Indeed, the Scottish research team found that spatial intelligence predicted a male's chances of successfully reproducing when playing 'sneaker', but not when playing 'guarder' (the researchers got each male to take a turn at each role, by introducing into the tank bigger and smaller rivals respectively). It would be a stretch to argue that human males' superior spatial abilities have a similar explanation (remembering where the most fertile women live!), but some have argued that spatial abilities were important for hunting, and hence ensuring that early man's partner and children didn't starve. Let's just hope that he didn't serve up mussels infused with a special fish sauce.

A Fishy Tale: You Found the Word

Cheat!! There was no word to find. Now turn to page 27 to learn about cheating in the animal kingdom.

A Fishy Tale: You Didn't Find the Word

There was no word to find; this wordsearch was a test of your honesty, which you passed with flying colours. Congratulations! Now turn to page 27 to learn about cheating in the animal kingdom.

ANSWER

Sorry about the skulduggery there. Ironically, perhaps, the only way I could find out whether or not you were tempted to cheat (albeit in a very low-stakes contest!) was to arrange a little deception of my own.

Do other animals cheat or try to deceive one another? The question is a difficult one to answer, because what might look like 'cheating' to us is just doing what comes naturally to the animal in question: non-human species don't have 'rules' for 'cheats' to 'break'.

Consider, for example, the humble cleaner fish. The deal is that the cleaner fish eats the parasites of its client, the reef fish. Both parties benefit from this arrangement: the reef fish gets to be clean, and the cleaner fish gets a good meal. However, cleaner fish often renege on this agreement, preferring instead to eat the tasty mucus secreted by the reef fish (yum!). Cleaner fish, in turn, punish these cheats by either chasing them aggressively or leaving them for a more reliable partner. This is often successful: an experimental study found that when cleaner fish are punished in this way, they reluctantly go back to their less preferred food source.

Now, as the paragraph above shows, it is perfectly possible to present this story as a kind of morality play: two fish make an agreement, one tries to cheat, is punished and finally sees the error of his ways (in fact, this sounds rather like the synopsis of the latest Disney–Pixar blockbuster). But it is equally possible to present this story in terms of simple associative learning, with no talk of cheating or punishment (*I do X, I get food; I do Y, I get better food; I do Z, I get no food*). In just the same way, reports of primates keeping quiet about the location of food (rather than giving the usual call) or even leading others in the wrong direction are not necessarily evidence of deception per se. The animal in question may have simply learned that, when he does such and such, he somehow ends up with more food, but has no idea why.

That said, a recent experimental study in which chimpanzees competed with a human experimenter for a food reward provides evidence that they are capable of at least some level of deception. A few pieces of banana were set out in front of the experimenter, who, if she spotted the chimpanzee approaching, quickly grabbed them for herself. The chimpanzees had a choice of two routes to the food: one involved a barrier, meaning that they could get to the food without being seen, the other a clear window. The cunning chimps preferred to take the hidden route, even when this was a long way around. Importantly, they were just as likely to do this at the start of the study as at the end. This suggests that chimps were not just learning a pattern of behaviour (*I do X, I get food; I do Y, I get no food*), but actually set out to 'deceive' their human competitor from the outset. Of course, it would be meaningless to say that the chimps were doing something morally wrong (wouldn't it?), but they do seem to understand that it is often possible to gain an advantage by hiding one's intentions. This certainly looks a lot more like 'cheating' than anything of which cleaner fish are capable.

Are You Big-Headed?

We now switch from fishy tales to fishy heads. But first, are you big-headed? (I'm not asking about ego here; I'm asking literally about the size of your head.) To find out, simply wrap a tape measure around your head and measure the circumference, making sure to measure at the widest point possible. This might take a few goes – just keep trying until you're fairly sure you've got the biggest measurement you possibly can (without cheating of course!). When you've got a measurement you're happy with, follow the steps below to work out the average head circumference for someone of your height (head size and height are – of course – related; think how weird the head of a strapping six-footer would look on someone only five feet tall). You can then see if your actual head is bigger or smaller than the average for someone of your height.

- First, take your height in centimetres, and multiply it by 0.087.
- Now add 42.4 if you are male, or 41.02 if you are female.

E.g., I am 5 foot 9 and a half, which is roughly 176.5cm

$$176.5 \times 0.087 = 15.36$$
$$15.36 + 42.2 = 57.75$$

So average head circumference for a man of my height is 57.75cm.

Once you've calculated the average circumference for someone of your height, you can figure out whether your head is on the large or small side. Roughly speaking, if your actual head circumference is 2cm bigger or 2cm smaller than the average for your height, then you are in the top 90% of big heads or the top 90% of small heads, respectively.

Why does this matter?

ANSWER

Ever since our species figured out that intelligence is 'housed' somewhere in the head we have wondered whether a bigger head equals greater intelligence. By the early 1900s, following pioneering work by Charles Darwin's cousin Sir Francis Galton, a positive relationship between head size and intelligence was pretty much taken for granted. Then along came the Nazis, who used racial differences in head size as a way to justify their racist ideologies. As a result, craniometry – the technical term for the science of head measurement – became socially unacceptable, and research ground to a halt.

In the modern era, studies using magnetic resonance imaging (MRI) and computed tomography (CT) – which allow us to measure the size of the *brain*, rather than just the skull – have largely confirmed the link with intelligence. Reviews published in 2005 and 2009 put the magnitude of the relationship between brain size and intelligence at somewhere between 0.3 and 0.4 on a 0–1 scale.* Even the rough-and-ready tape-measure method gives a correlation of around 0.2 – small, but easily statistically significant (the concept of statistical significance is explained in the section Parrotnormal Activity). Interestingly, the relationship between brain size and intelligence seems to be stronger for females than males, and for adults than children.

What about other animals: are species with bigger brains smarter? Of course, it's difficult to measure intelligence in the same way across very different species. But, in general, yes, as long as you measure brain size relative to the overall size of the animal, and correct for diminishing returns (e.g., a 10 per cent increase

---

* (Scientists express the size of the relationship between two variables – such as brain size and intelligence – using a number called a *correlation coefficient*, which ranges from 0 to 1. Zero indicates no relationship whatsoever, and 1 indicates a perfect relationship: e.g., if you know a person's brain size, you can predict his or her intelligence perfectly. So, a correlation of 0.3–0.4 is not *huge*, but it is certainly not to be sniffed at.)

in brain size will make more difference to a tiny animal than to a huge animal).

The link between brain size and intelligence leaves us with an evolutionary puzzle: if bigger is better, why aren't brain sizes rocketing? A long-standing answer is the *expensive-tissue hypothesis*. Compared with common-or-garden skin and bone tissue, brain tissue requires a huge amount of energy, and resources are limited. While in principle a huge brain would be useful, in practice few animals could consume enough calories to keep it going. One way to test this idea is to investigate whether animals that have evolved bigger brains are forced to compensate by having a smaller heart, kidney, liver or gut – all tissue that is relatively expensive (in most cases, even more expensive than brain tissue).

This is where the fish heads come in. Perhaps the most direct test yet of the expensive-tissue hypothesis comes from a study in which Swedish researchers selectively bred guppies with particularly large and small brains (relative to body size). The big-brained guppies that emerged as a result were not only more intelligent (i.e., they were better at learning that a card with four symbols, but not two symbols, led to a food reward) but also had smaller guts than their small-brained, dimmer tank-mates. Interestingly, these guppy geniuses also produced fewer children, suggesting that – exactly as predicted by the expensive-tissue hypothesis – evolving a big brain involves sacrificing other parts of the body, including those that contribute to the ability to reproduce.

Now, I hope for your sake that you have a big brain and a small gut, as next up is probably the trickiest test in the book – possibly, in fact, the most devilish puzzle I've ever come across.

Goatbusters

Shown below are three doors. Behind one is a car, behind each of the other two, a goat. Your job is to try to win the car by opening the correct door. Go ahead, pick one.

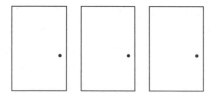

After you've picked a door, I open one of the other two doors, revealing a goat. Then, before opening your door, I give you a choice: you can stick with the door that you originally picked or switch to the remaining, unopened door. Do you stick or switch? (If you're already familiar with this problem, feel free to skip straight to the answer; but unless you're an expert in animal behaviour, I promise you'll still learn something new.)

ANSWER

The surprising answer is that it's always better to switch. The advantage of switching has been proven both mathematically (i.e., on paper) and empirically (i.e., by getting people to play the game a bunch of times and showing that they win the car much more often if they always switch than always stick).

This 'Monty Hall Problem' (named after the host of a game show that featured this puzzle as its finale) has become pretty famous in recent years, so it's quite possible that you already knew the answer. That said, it's such a devilish and counter-intuitive puzzle that even many people who *do* know the answer still don't quite believe it in their bones, or understand exactly why it's right. Indeed, when Marilyn vos Savant, a former holder of the title of World's Highest IQ, pointed out the answer, she was subjected to a deluge of 'corrections', including from maths professors and other 'experts' in probability theory.

There are many different ways to understand why switching is best, but the simplest is as follows. One-third of the time your initial choice will be the 'car' door, so you would be better off sticking. Two-thirds of the time your initial choice will be a 'goat' door, so you're better off switching. Therefore, you're twice as likely to win if you switch than if you stick. Not that people do. In both the original TV game show (*Let's Make a Deal*) and subsequent laboratory versions, people overwhelmingly choose to stick with their original door.

At this point you're probably wondering what all this has to do with animals. After all, the only ones we've met so far have been booby prizes. The answer is that a pair of enterprising scientists conducted a study to find out how pigeons would fare at the Monty Hall Problem. The pigeon version uses three 'keys': light-up buttons that, as the pigeons have already learned, sometimes give out food rewards when pecked. At the start of the game all three keys are lit up white. Once the pigeon has pecked one key (i.e., chosen its initial 'door'), both this key and one of the other two keys are lit up green, to offer the pigeon the stick-or-switch. The pigeon then pecks one or other of these keys, and either does or does not receive a food reward, depending on which had been pre-designated as the 'prize' key. Of course, pigeons cannot be given instructions. So, unlike the human game-show participants, pigeons are given hundreds of runs through the game.

The results are quite astounding. After thirty days of training (with, in most cases, a hundred runs per day) pigeons had learned to switch rather than stick on over 96 per cent of trials. You might object that humans could learn to switch if they were given such extensive training, but this does not seem to be the case. When humans (well, university students) completed a close analogue of the pigeon version, using a touchscreen rather than peckable keys, they were still no more likely to switch than stick after 200 training trials. (Although this was still a fraction of the number of trials that our feathered friends received, if you haven't figured out the game after 200 goes, you're probably not about to.)

Why are we so bad at this task? One possible explanation is that humans are subject to a phenomenon known as *loss aversion* (as I discussed in my last book, *Psy-Q*). Psychologically speaking, losing something that you already have is much worse (in fact, roughly twice as bad) than failing to gain that same thing. While, from a mathematical perspective, loss aversion is irrational, from an evolutionary perspective it makes good sense: if you have a food stash that will get you through the winter, it is much better to hang on to what you've got than to go out seeking more. It would be pure madness to view failing to double the size of your stash as *just as bad* as losing what you already have, which is what a complete lack of loss aversion would entail. Thus at least part of the reason that people tend to stick with their initial choice is that they assume – probably correctly – that they would feel much worse after switching from a winning door to a losing one than after picking a losing door and sticking with it. We are also probably influenced by the fact that, in everyday life, it is generally considered much more noble and courageous to 'stick to one's guns' than to 'flip-flop' (even if the initial choice is essentially random).

A third possible reason for the fact that humans are outperformed by pigeons on this problem is that we are simply too clever for our own good. Most people try to calculate the probabilities of winning under both the stick and switch scenarios and, having concluded – incorrectly – that it is 50/50 either way, figure that they may as well stay with their initial choice (presumably due to loss aversion and the desire to stick to one's guns). Like the misguided mathematicians who plagued poor Marilyn vos Savant, we are so sure of our own misguided abilities at probability calculations that we refuse to abandon them, even in the face of 200 trials' worth of counter-evidence. Pigeons, on the other hand, have none of these mental blind spots, and so just learn the probabilities on the job, rather than trying to figure them out beforehand.

Whatever the reason, one thing is clear: for pigeons, the fact that they can beat us humans at a game we invented is certainly a real coo.

Pigeon's Treat

So, it's Pigeon 1–0 People in this avian-versus-ape showdown. Let's see if you can recover some dignity for our species by taking on these brainy birds in four more tests.

Phones for You. You've just saved up to buy a fancy new phone. You had to really put in the hours in your part-time job (which you hate with a passion), but it was worth it. You place your order online, and the phone arrives first thing in the morning. That afternoon, an identical phone arrives. You contact the company, who – after keeping you on hold for an hour, and failing to phone you back twice – say that their system can't process a return and, in fact, you'd be doing the stressed-out guy in the call centre a favour if you just kept the phone and didn't call back. You agree, and decide to treat your brother, whose birthday is coming up, and whose current phone is all but unusable. But which of the two still-shrink-wrapped phones do you give him?

☐ The first one   ☐ The second one

Band Aid. Three months ago you bought a £150 advance ticket to see one of your favourite bands. Then yesterday, your *favourite* band announced a new tour, and you quickly snapped up a £100 ticket. In your excitement, you forgot to check the dates and – you guessed it – the shows are on the same night. You can't sell either ticket: both of these bands are so obscure that their gigs never sell out, and everyone you know hates them. Which do you attend?

☐ £150 gig   ☐ £100 gig

Don't be a Mug. You really want to buy some cool vintage coffee mugs, and the more mugs the better (you hate washing up, and have big cupboards). You go to a flea market. One seller has a box of twenty mugs, though three have nasty chips and two are missing handles. Another seller is offering, for the same price, a box of twelve intact mugs. You can't buy both because – oh, I don't know – the two sellers hate each other, and each won't deal with you if you've bought off the other. From whom do you buy your mugs?

☐ First seller ☐ Second seller

A Taxi Attacks. There are two taxi companies in your town: Crimson Cabs (who get 85 per cent of jobs) and Turquoise Taxis (who get only 15 per cent, perhaps because people don't like the colour of their cars). A taxi was involved in a hit-and-run incident. In court a witness said that the car was turquoise, and the prosecution provided evidence that – under the same lighting conditions as the night in question – this witness can correctly tell apart crimson and turquoise cars 80 per cent of the time. What is the probability that the Turquoise Taxi driver is responsible for the hit and run?

☐ per cent

ANSWER

Phones for You. Well, there's no 'right' answer here; the whole point is that it makes no difference. But, if you're honest, and this happened for real, I bet you'd give him the second, free one, wouldn't you? If so, you are showing a *justification of effort* effect: you value things that you have to work hard for much more than (identical!) things that come cheap or for free. But in cases such as this one this is a logical fallacy: it makes absolutely no difference which phone you give away and which you keep.

Pigeons show the same fallacy. Suppose pigeons are trained that both a red key and a green key give two seconds of access to grain when pecked. The clever part is that, in order to access the red key, the pigeons need to give one peck on a white key; but in order to access the green key, they need to give twenty pecks on the white key. Finally, pigeons are given a free choice between the red and green key, without needing to peck on the white key at all. Which one do they prefer? Yes, the one that they usually had to work hard to get, even though – just as with the two phones – both are exactly as good.

Band Aid. This time, there is a right answer: you should just go and see your favourite band. If you decide to go to see the other band, you are showing a **sunk cost** effect. Having already 'sunk' a lot of money into the £150 ticket, you can't bear to waste it. Again, this is a fallacy. The past is irrelevant; that £150 is gone for ever whatever you do, so just go to the gig you'll prefer.

Again pigeons (and also starlings) show the same fallacy. Suppose a pigeon has already pecked ten times on a green key. Now, in order to earn his food reward, he must give either another twenty pecks on the green key or ten 'new' pecks on a red key. Even though he could save himself ten pecks worth of effort by switching to the red key, the pigeon prefers to stick with the green key, so as not to 'waste' the ten pecks that he has already 'sunk' into this key.

Don't be a Mug. The first seller is, in effect, offering fifteen mugs, whereas the second is offering twelve mugs for the same price. You would be crazy to go with the second seller, just to save yourself the minor inconvenience of carrying home and throwing away three mugs. If you did so, you are showing a *less is more* effect (i.e., thinking you're getting *more* value by getting *less* – all right, *fewer* – pristine mugs). Again this is a fallacy. Less is *not* more. More is more. The fallacy arises because people tend to average over the whole set when making their judgement. For example, in one study, participants guessed that a hamburger had 734 calories, but that a hamburger plus three sticks of celery (the saddest Happy Meal I've ever seen) had only 619 calories (and, no, they didn't think that eating a stick of celery burns 38⅓ calories).

You guessed it: pigeons again show the same fallacy. When given the choice between a pea alone and a pea plus a piece of milo (a relatively unappetising grain), pigeons choose the pea (unless they have been starved beforehand, in which case they go for the meal deal). Similarly, dogs will choose a piece of cheese over a piece of cheese plus a bonus carrot, and macaques will choose a grape over a grape plus a bonus green bean. It's not that they hate the pea, carrot or green bean – they'll eat it if that's all that's on offer – it's just that pigeons, dogs and monkeys, like humans, think that less is more.

A Taxi Attacks. If you said 80 per cent, you are showing an effect of *base-rate neglect*. You're neglecting the base-rate of crimson vs turquoise taxis on the road at any one time (85 per cent vs 15 per cent). Before the witness showed up, the chances of a crimson taxi being to blame were 85 per cent, not 50/50. The correct answer is 41 per cent. Let's work out the probability of all four possibilities (though we only really care about the last two):

- Witness correctly identifies car as crimson (80 per cent success rate × 85 per cent crimson cars = 68 per cent)
- Witness incorrectly identifies turquoise car as crimson (20 per cent error rate × 15 per cent turquoise cars = 3 per cent)
- Witness correctly identifies car as turquoise (80 per cent success rate × 15 per cent turquoise cars = 12 per cent)
- Witness incorrectly identifies crimson car as turquoise (20 per cent error rate × 85 per cent crimson cars = 17 per cent)

So, overall, forgetting about whether he's right or not, there is a 29 per cent chance that the witness will *say* that the car is turquoise (12 per cent + 17 per cent). But, as we've already seen, he was right only for the 12 per cent. So to get the overall probability that the witness was correct to say that the car was turquoise we have to divide 12 per cent by 29 per cent, giving us 41 per cent. If that's too complicated, here's the bottom line: the witness doesn't get it wrong *that* often (just 20 per cent of the time), but Turquoise Taxis get a job even *less* often (just 15 per cent of the time). So it's more likely that the witness is wrong than that a turquoise car managed to secure this (ultimately ill-fated) job.

If you thought that was difficult, imagine trying to come up with a pigeon version. Here's what the researchers did. In the first part of the study, a 'sample' key lights up either red or green, and the pigeon wins its treat by (after an enforced delay of a few seconds) pecking either another red key or another green key – whichever matches the colour of the sample key. In the second part of the study the sample key lights up either red or white, and the pigeon wins its treat by (again, after an enforced delay) pecking either a key with a line (if the sample was red) or a key with a dot (if the sample was white). The important thing to bear in mind here is that the sample key is red (50 per cent of the time) more often than it is either green (25 per cent) or white (25 per cent).

So, how did they get on? When the sample was red, the pigeons almost always correctly pecked the red key (rather than the green key) or the line (rather than the dot). But when the sample was green or white, pigeons got it wrong more often than not, pecking the key that follows a red sample (i.e., either red or the line). Why? Base-rate neglect. Remember, the sample key is red (50 per cent) twice as often as it is either green (25 per cent) or white (25 per cent). So, together, the responses associated with a red sample (red key + line key) lead to food twice as often as those associated with the green sample (green key) or white sample (dot key). Thus pigeons choose the 'red sample' responses because these are more likely to lead to food *overall*, ignoring the fact –

for *any particular choice* – the 'red sample' response has only a 50 per cent chance of being correct (this is the 'base rate' that they're neglecting).

What was your score overall? Did you beat the pigeons?

		Pigeons		You	
Problem	Phenomenon	Got it right	Fell for the fallacy	Got it right	Fell for the fallacy
Goatbusters	Monty Hall Problem	✓			
Phones for You	Justification of effort		✓		
Band Aid	Sunk costs		✓		
Don't be a Mug	Less is more		✓		
A Taxi Attacks	Base-rate neglect		✓		

Probably not. The point of these studies was to show that pigeons show the same logical fallacies that are already known to be widespread in humans. Why do we show these fallacies? Nobody knows for certain, but Thomas Zentall, who published a review paper summarising these studies, has some suggestions.

If an animal places more value on food that it has had to work hard for (*justification of effort*), then that may motivate it to persist longer when looking for food. *Sunk cost* effects may arise from the fact that, once you've got a food source you're relatively happy with, moving seems unnecessarily risky (see Should I Stay or Should I Go? and More Monkeyconomics), and this conservatism spills over into choices where there is in fact no such risk. *Less is more effects* look puzzling to humans, but remember that most animals can't count (or, at least, not very well). This means that, often, the best they can do is judge the overall average quality of two rival sources of (mixed) food, rather than work it out piece by piece. And if counting is difficult, then calculating probabilities is nigh on impossible, for pigeons and humans alike; hence the phenomenon of base-rate neglect.

If you did manage to beat the pigeons, congratulations! But don't rest on your laurels. It's now time to take on a much more intelligent, more closely related, species . . .

MMMMonkeying Around

Yep, it's those crafty capuchins, who are probably already beating you 1–0 from An Expensive Cappuccino. Here's your chance to even up the score.

Below are two M&Ms. You can choose either the red or the blue one:

Now, here's a yellow M&M. You can either have this one, or the remaining red or blue one (i.e., the one that you *didn't* choose last time).

Chosen? Now turn the page to find out what capuchins did when given the same choice.

ANSWER

In the capuchin version, conducted at the prestigious Yale University, monkeys chose the new M&M rather than the one that they had previously rejected on 60 per cent of trials. Now 60 per cent may not sound like a lot, but it is significantly greater than the 50 per cent that would be expected if they were choosing at random. And, after all, why not just choose at random? All M&Ms taste the same, and the researchers checked beforehand that, when given a free choice between M&Ms of all three colours, the capuchins showed no preference.

But unlike in An Expensive Cappuccino, where the monkeys come out on top, this is a test for which capuchins and humans (or, at least, children) are on a par. When the researchers did the same study with human four-year-olds, using stickers rather than M&Ms, they found that children also chose the new item over the previously rejected one around 60 per cent of the time (the experimenters made sure to use stickers that each child had rated beforehand as equally desirable). The mystery deepens still further with the finding that, if the *experimenter* made the first choice, neither monkeys nor children preferred the new M&M or sticker over the old one. (In fact, the monkeys preferred the old one, perhaps because they thought that – in the first part of the procedure – the experimenter was deliberately keeping the tastier treat for himself.) There seems to be something special about *rejecting* one of the two options that makes it become less desirable. What's more, the fact that monkeys do it too suggests that this downplaying of rejected alternatives doesn't involve deliberate reasoning but is somehow automatic, possibly even genetically hard-wired.

What's going on? How is it that experimenters are able to monkey around with our preferences in this way? Psychologists explain these types of phenomena as arising from *cognitive dissonance*, an uncomfortable nagging feeling that we get when our actions (e.g., choosing one sticker over another) conflict with our beliefs ('both of those stickers are equally good'). Since this feeling is uncomfortable, we do what we can to get rid of it. Since we can't change our actions (choosing both stickers at Stage 1 is against the rules of the game), the only way to escape from cognitive dissonance is to change our beliefs ('actually, the sticker I rejected isn't quite as good as the one I chose'). Since the new sticker hasn't suffered this put-down, it then starts to look like the more attractive option.

Although this study looked at capuchins and children, adults aren't off the hook, as we can see from the 1956 study that kick-started this line of research. In a *Generation Game*-style set-up that sounds slightly sexist to modern ears, female college students were shown a set of domestic appliances (including a coffee-maker, an electric sandwich grill and a toaster) and asked to rate them for desirability. Next, each participant was shown a pair of items and invited to choose one to take home as payment for participating in the experiment. When asked to rate the same items again, participants increased the rating of the item that they had chosen, and decreased the rating of the one they had rejected. (The experimenter then spoiled everyone's fun by revealing that actually nobody would get to keep their prize, causing one participant to burst into tears.)

These *preference-reversals* aren't just an experimental curiosity; they turn up in the real world too. For example, an estate agent who wants you to buy House C (which is really a bit beyond your budget) rather than House B (which is well within it) may first offer you a choice between House A and House B, in the hope that you will reject B, thus making you less likely to choose it when House C 'comes on the market' (having really been there all along). There may also be a similar effect at play with supermarket own-brand ranges. Perhaps, in a perverse sort of way, having rejected the standard, mid-range product as poor value (compared to the no-frills 'essential' or 'basic' version), you are then more likely to reject it in favour of the last option you consider: the most expensive, 'best' or 'finest' version (particularly if you throw a bit of price-equals-quality reasoning – see An Expensive Cappuccino – into the mix). Indeed, who's to say that you aren't the unwitting victim of a similar ruse every time you're handed a menu or a wine list?

Economists, of course, are wise to this game, which brings us to a probably apocryphal exchange that I first came across in Robert Frank's book *The Economic Naturalist*:

CUSTOMER: What sandwiches have you got?
WAITER: Chicken salad and roast beef.
CUSTOMER: OK, I'll have the roast beef please.
WAITER: Oh yes, we also have tuna.
CUSTOMER: Right. In that case, I'll have the chicken salad.

More Monkeyconomics

We'll move on from capuchins in a minute, I promise. But first, here is one final example of monkeyconomics.

1. Would you prefer . . .
   (a) to win £50 for sure **OR** ☐
   (b) a 50 per cent chance to win £100 (50 per cent chance to win nothing) ☐
2. Would you prefer . . .
   (a) to lose £50 for sure **OR** ☐
   (b) a 50 per cent chance to lose nothing (50 per cent chance to lose £100) ☐

ANSWER

Did you choose 1a and 2b? If so, you are again showing the same pattern as capuchins.

In this set up, as in An Expensive Cappuccino, the monkeys were given tokens that could be exchanged for food, and could choose which of two experimenters to deal with.

In the first part of the study (like your Question 1), Experimenter A (the 'sure thing') started out displaying one piece of apple but, after receiving the token, always gave the monkey this piece and a bonus piece. Experimenter B (the 'gamble') also started out displaying one piece of apple but, after receiving the token, gave the monkey either just this one piece or this piece and *two* bonus pieces (each half of the time). Here the capuchins preferred (a) the 'sure thing', just as most humans choose (a) Win £50 for sure.

In the second part of the study (like your Question 2), Experimenter A (the 'sure thing') started out displaying three pieces of apple but, after receiving the token, always binned one piece and gave the monkey only two. Experimenter B (the 'gamble') also started out displaying three pieces of apple but, after receiving the token, gave the monkey either all three pieces or just one (each half of the time). Here the capuchins preferred (b) the 'gamble', just as most humans choose (b) 50 per cent chance to lose nothing.

This may not seem that surprising at first, but look a little closer. In both parts of the study the monkeys had a choice between (a) a sure thing, getting two pieces of apple every time and (b) a gamble, getting either one piece or three pieces. So why did they chose the sure thing in the first part of the study and the gamble in the second? It's all to do with *framing*. Remember that, in the first part of the study, both experimenters started out displaying just one piece of apple, while, in the second part, both experimenters started out displaying three pieces. As a consequence, choosing the sure thing looks like a *guaranteed gain* in the first part of the study (two pieces instead of the anticipated one piece) but a *guaranteed loss* in the second part of the study (two pieces instead of the anticipated three).

When choices are framed as potential *gains*, monkeys are *risk-averse* ('Right, I'm getting two pieces of juicy apple here; let's not risk losing one by being greedy'). But when choices are framed as potential losses, monkeys are *risk-seeking* ('Dammit, I really wanted those two pieces of apple, but now it looks like I've already lost one. Sod it, let's go for the gamble; then I might win it back').

Silly monkeys, eh? Except that – yes – we humans do exactly the same thing. If, like most people, you chose (a) the sure thing when you stood to win money and (b) the gamble when you stood to lose money, you are no different from the capuchins in this experiment. Although Question 1 was framed in terms of winning money and Question 2 in terms of losing money, the choice in both cases is exactly the same: do you take a gamble with a 50/50 chance of leaving you £50 better off than you would have been with the safe choice? If you remain unconvinced, try this version:

1. You have just won £100. Now would you prefer . . .
   (a) to win an extra £50 for sure OR ☐
   (b) a 50 per cent chance to win an extra £100 (50 per cent chance to win no extra money)? ☐
2. You have just won £200. Now would you prefer . . .
   (a) to pay back £50 for sure OR ☐
   (b) a 50 per cent chance to pay back nothing (50 per cent chance to pay back £100)? ☐

Again, almost everybody prefers (a) the safe bet when the question is framed in terms of winning (Q1) but (b) the gamble when the question is framed in terms of losses (Q2). But in this version it is even easier to see that the two questions are exactly equivalent. In each case you can either (a) walk away with £150 or (b) gamble, and walk away with either £100 or £200.

As for the *preference-reversals* we met in the last section, the fact that these framing effects are found in both monkeys and humans suggests that they may have some common evolutionary origin. But this leads to a puzzle: how could it be an advantage, in terms of natural selection, to have biases that fly in the face of logic (i.e., *risk-aversion* when faced with gains and *risk-seeking* when faced with equivalent losses)?

Presumably the answer is that, while these biases are illogical in carefully controlled experimental set-ups, they may be highly adaptive in the wild. For example, if you have just enough food to survive, it makes perfect sense to be risk-averse and carefully guard what you have, rather than gamble by hiding your food and going off hunting. On the other hand, if you have just lost most of your stash and are about to starve to death, it makes perfect sense to roll the dice and go off hunting. Similarly, while downplaying the attractiveness of an M&M or toaster that you have just rejected may be illogical in a lab experiment, downplaying the attractiveness of a job or a partner that you have just rejected may be crucial for your sanity.

So let's raise a glass to our 'illogical' primate biases. Three cheers for monkeyconomics!

Ape, Man United

We have just seen (in An Expensive Cappuccino, MMMMonkeying Around and More Monkeyconomics) how our primate heritage has left us with some faulty thinking biases, but what can it tell us about Premier League football?

Of the 92 English clubs (Premier League, Championship, League 1 and League 2), guess how many play in each of the following colours? Only home kits are counted, and trims and other minor details are ignored (two colours are shown only for striped or hooped kits where both colours are roughly equally dominant).

Black and White	
Blue (light and/or dark)	
Blue and White	
White	
Green and White	
Grey and White	
Red and White	
Red	
Red and Black	
Red and Blue	
Claret	
Orange (including Gold/Amber)	
Orange and Claret	
Orange and Black	
Yellow	

Here we turn to ape man extraordinaire Desmond Morris. Morris made headlines worldwide with his 1967 book *The Naked Ape,* which sought to explain key aspects of human behaviour in terms of our evolutionary origins. (The book's title refers to the fact that, out of almost 200 different primate species, humans are the only ones not covered with thick body hair.) Morris developed this theme in *The Human Zoo,* which drew parallels between city-dwelling humans and animals in captivity, and *Manwatching: A Field Guide to Human Behaviour.* He then turned his attention to football (OK, 'soccer') with *The Soccer Tribe: A Companion Volume to Manwatching,* which analysed the beautiful game from his by now trademark zoological perspective (sample chapter headings: 'The Soccer Match as a Ritual Hunt'; 'The Challenge of Playing on Rival Territory'; 'The Triumph Displays'; 'Tribal Chants'). This section draws its inspiration from Morris's chapter on 'Tribal Colours'.

Here are the answers: (a) in the same order as the previous page and (b) in order of popularity of colours.

Black and White	2	Red	21
Blue (light and/or dark)	17	Blue (light and/or dark)	17
Blue and White	10	White	12
White	12	Blue and White	10
Green and White	2	Red and White	9
Grey and White	1	Orange (including Gold/ Amber)	5
Red and White	9	Claret	4
Red	21	Yellow	4
Red and Black	1	Black and White	2
Red and Blue	1	Green and White	2
Claret	4	Orange and Black	2
Orange (including Gold/ Amber)	5	Grey and White	1
Orange and Claret	1	Red and Black	1
Orange and Black	2	Red and Blue	1
Yellow	4	Orange and Claret	1

Morris argues that the popularity of highly conspicuous colours (nobody plays in brown, grey or pastel shades) is due to the fact that they give their wearers a psychological advantage by demonstrating their fearlessness. He draws parallels with the warning coloration of poisonous snakes (orange and black), wasps (yellow and black) and skunks (black and white), and even goes so far as to suggest that 'venomous colour patterns of this type give the wearers' enemies an unconscious sensation that they may be "stung" or "poisoned" by too-close contact' (p. 205). Yes, he's talking about rival footballers.

Why, then, aren't yellow and orange more popular? Morris argues that other psychological associations are at work. Yellow is associated with cowardice, and orange is seen as an intermediate colour that can't make up its mind (notice how the vast majority of kits use primary colours). Red, on the other hand, symbolises 'blood, energy, life, force, power and intensity'. The popularity of blue is harder to explain, but Morris suggests it may have a calming effect on teammates. Another reason why many teams choose blue (or white) is presumably to contrast with local rivals who wear red, ensuring that both teams can wear their first-choice kit on derby day (Man Utd vs Man City; Liverpool vs Everton; Arsenal vs Spurs). White is also perhaps more popular than one might expect, perhaps because of its associations with heroism.

That's all very well, but can we put any of this to the test? Perhaps. Morris argues that, owing to its symbolic associations, red is 'the perfect colour for any sporting team and it is difficult to understand why, in some countries, it is not even more popular than it already is'. If he is right, then we might expect to see red predominate among the English champions. In fact, although red is somewhat over-represented among the title winners, it's not by much. In general, the colours of the title winners show a similar distribution to that seen across the 92 clubs, with 52 red champions (Manchester United, Liverpool, Arsenal and Nottingham Forrest), 22 blue (Chelsea, Everton, Manchester City, Portsmouth, hooray! – my team – Ipswich Town, and – double hooray! – against all odds, Leicester City), 11 blue and white (Huddersfield Town, Sheffield Wednesday, West Bromwich Albion and Blackburn Rovers), 9 white (Leeds United, Tottenham Hotspur, Preston North End and Derby County), 7 red and white (Sunderland and Sheffield United), 9 claret (Aston Villa and Burnley), 4 black and white (Newcastle United) and 3 orange (or 'old gold', as Wolverhampton Wanderers insist on calling it).

And even if red teams do a bit better than one might expect, this is due mainly to just three teams: Man Utd, Liverpool and Arsenal, with 20, 18 and 13 titles respectively. Morris's analysis breaks down if we look at, say, Spain, where – with the odd exception – Real Madrid (white, 32 titles) and Barcelona (red and blue, 24 titles) pass the title back and forth. Although Bayern Munich (red, 25 titles) are hugely dominant in Germany, France is similarly dominated by the pair of Marseille (white) and Saint-Étienne (green!), with 10 titles each. Italy at least provides some support for the idea of 'warning coloration', with Juventus, in their skunk-like kit, winning 32 titles; but the rival red-and-black and blue-and-black teams of Milan (A.C. Milan and Internazionale), with 18 each, have literally nothing between them.

So, in terms of league titles, Morris's account of tribal colours is looking a bit, well, washed out. But if you dig a little deeper, and look at win percentage for home games only (where teams are always wearing their first-choice kit), it turns out that red teams do indeed show better performance than would be expected by chance. Crucially, the 'red' teams did not outperform the others in terms of win percentage in away games, for which they often wear a non-red second kit. This is important, as it suggests that there is an advantage to be gained from actually *wearing* red, as opposed to just being Man Utd, Liverpool or Arsenal. That said, this analysis is still not watertight: it seems likely that these three big clubs, with their big stadiums and large, enthusiastic fan-bases, will enjoy more of a home advantage than smaller clubs. Again, the acid test will be to see if the same pattern is observed for other countries whose leagues are not dominated by a small, red-wearing elite.

Although goalkeepers seldom wear red, the evidence that they would be better off doing so is actually pretty convincing. One recent study had the same players take penalty kicks against a goalkeeper wearing different coloured shirts (two different goal-keepers were used, but each changed his shirt after every kick, so wore each colour an equal number of times). Although the penalty takers rated themselves as equally confident in each scenario, they actually scored significantly fewer when the goalkeeper was wearing red than when he was wearing blue or green.

Thus, while it is easy to mock, Morris's claim that certain co-lours confer a psychological advantage seems to be supported. But let's not take it too far: in terms of domestic dominance, Juventus are arguably the most successful team in Europe. But their skunk-like warning coloration sounds a lot less fearsome when you learn that it was borrowed from Notts County.

Having a Mare (or not)

With apologies to Magpies fans, it is fair to say that Notts County are hardly the sexiest team in the league. But here's something that should get your pulse racing.

Don't just take my word for it, let's check. Before reading on, please take your pulse.

Now carefully study whichever of the following stock photos you prefer (or, if you can't decide, both). While hardly pornographic, at least one of these pictures should get your pulse racing a little (if not, I'm sure you'll be able to find one that does easily enough). Take your pulse again to check.

Once your pulse has increased, let out a big sigh. What did you find?

ANSWER

Did your pulse decrease when you sighed? This was the finding when researchers did the same test with horses. Fifteen stallions were fitted with heart rate monitors and presented with an attractive mare (not a picture, a real one – but one well out of reach). The stallions' heart rates shot up as they got all hot and bothered, but went back to normal as they whinnied. Just as with humans, vocalising seems to reduce tension and excitement. So next time you find yourself getting a little hot under the collar (for whatever reason), try a big sigh. You can probably get away with this in public, as long as you do it under your breath. But, for God's sake, please don't whinny.

A Stable Personality?

If these stallions got really desperate, they might settle for mating with a female donkey. Such unions are usually unsuccessful, but if the jenny did manage to give birth, we might want to submit the poor hinny for psychological testing.* Fortunately, researchers have developed a genuine scientifically validated personality test.

Crazy as this may sound, it's not even unusual. Such tests have been developed and validated for hyenas, pigs, rats, guppies and octopuses, for chimpanzees, gorillas and monkeys (rhesus and vervet), and – inevitably – cats and dogs. At first glance, animal personality tests may seem like one big joke. But, as perhaps the leading authority on animal personalities, Sam – I kid you not – Gosling, points out, 'there is nothing in evolutionary theory to suggest that only physical traits are subject to selection pressures, and Darwin (1872) argued that emotions exist in both human and nonhuman animals'. For example, it is not difficult to see how a confident and outgoing animal might enjoy more success at the 'three *fs*' – feeding, fighting and fornicating – than a shy retiring type.

Gosling's review of nineteen animal personality studies concluded that three of the five traits commonly held to make up human personalities – **Extraversion, Neuroticism** and **Agreeableness** (defined below) – have close analogues across pretty much all of the species listed above, except that guppies and octopuses don't seem to vary in Agreeableness. (Well, have you ever met an octopus that you didn't like?) Of course, the manifestations of these traits differ widely across species: 'whereas the human scoring low on Extraversion stays at home on Saturday night, or tries to blend into a corner at a large party, the octopus scoring low on

---

* A *hinny* is the offspring of a male horse and a female donkey. The better-known *mule* is the offspring of a female horse and a male donkey.

Boldness stays in its protective den during feedings and attempts to hide itself by changing colour or releasing ink into the water'. A version of **Openness to Experience** is found mostly in other primates, but also hyenas, pigs, dogs and cats, although only chimps seem to vary in their **Conscientiousness**. Are there any personality dimensions that characterise individual animals but not humans? Perhaps. Animals vary systematically in their **Dominance**, which does not seem to be a stable personality trait in humans (for example, intellectuals typically enjoy dominance in the academic – but not the sporting – arena).

The 'Big Five' dimensions of human personality (from Gosling and John, 1999: 70):

Neuroticism (anxiety, depression, vulnerability to stress, moodiness)

Agreeableness (trust, tender-mindedness, cooperation, lack of aggression)

Extraversion (sociability, assertiveness, activity, positive emotions)

Openness to experience (ideas/intellect, imagination, creativity, curiosity)

Conscientiousness (deliberation, self-discipline, dutifulness, order)

Enough talk. It's time to compare your personality with that of your dog (or, if you don't have one, any dog that you know, or even your favourite fictional canine).

Overleaf are a number of characteristics that may or may not apply to you and your dog. For example, do you agree that you/your dog *like(s) to spend time with others*? Please circle the number in the relevant box below to indicate the extent to which **you agree or disagree with that statement**.

	You						Dog				
	Disagree strongly	Disagree a little	Neither agree nor disagree	Agree a little	Agree strongly		Disagree strongly	Disagree a little	Neither agree nor disagree	Agree a little	Agree strongly
1. Is talkative, vocal	1	2	3	4	5		1	2	3	4	5
2. Is disagreeable, difficult to please	5	4	3	2	1		5	4	3	2	1
3. Does things thoroughly	1	2	3	4	5		1	2	3	4	5
4. Is down, depressed, blue	5	4	3	2	1		5	4	3	2	1
5. Is original, comes up with new ways of doing things	1	2	3	4	5		1	2	3	4	5
6. Is reserved	5	4	3	2	1		5	4	3	2	1
7. Is helpful and unselfish	1	2	3	4	5		1	2	3	4	5
8. Can be somewhat careless	5	4	3	2	1		5	4	3	2	1
9. Is relaxed, handles stress well	1	2	3	4	5		1	2	3	4	5
10. Is curious about many different things	1	2	3	4	5		1	2	3	4	5
11. Is full of energy	1	2	3	4	5		1	2	3	4	5
12. Starts quarrels with others	5	4	3	2	1		5	4	3	2	1
13. Is a reliable person/dog	1	2	3	4	5		1	2	3	4	5
14. Can be tense	5	4	3	2	1		5	4	3	2	1
15. Appears contemplative, thoughtful	1	2	3	4	5		1	2	3	4	5
16. Shows a lot of enthusiasm	1	2	3	4	5		1	2	3	4	5
17. Has a forgiving nature	1	2	3	4	5		1	2	3	4	5
18. Tends to be disorganised	5	4	3	2	1		5	4	3	2	1
19. Worries a lot	5	4	3	2	1		5	4	3	2	1

	You					Dog				
	Disagree strongly	Disagree a little	Neither agree nor disagree	Agree a little	Agree strongly	Disagree strongly	Disagree a little	Neither agree nor disagree	Agree a little	Agree strongly
20. Is unimaginative, dull	5	4	3	2	1	5	4	3	2	1
21. Tends to be quiet	5	4	3	2	1	5	4	3	2	1
22. Is generally trusting	1	2	3	4	5	1	2	3	4	5
23. Tends to be lazy	5	4	3	2	1	5	4	3	2	1
24. Is emotionally stable, not easily upset	1	2	3	4	5	1	2	3	4	5
25. Is inventive, finds new ways to get his/her way	1	2	3	4	5	1	2	3	4	5
26. Has an assertive personality	1	2	3	4	5	1	2	3	4	5
27. Can be cold and aloof	5	4	3	2	1	5	4	3	2	1
28. Perseveres until the task is finished	1	2	3	4	5	1	2	3	4	5
29. Can be moody	5	4	3	2	1	5	4	3	2	1
30. Appreciates sensory experiences	1	2	3	4	5	1	2	3	4	5
31. Is sometimes shy, inhibited	5	4	3	2	1	5	4	3	2	1
32. Is considerate and kind	1	2	3	4	5	1	2	3	4	5
33. Does things efficiently	1	2	3	4	5	1	2	3	4	5
34. Remains calm in tense situations	1	2	3	4	5	1	2	3	4	5
35. Enjoys learning and doing new things	1	2	3	4	5	1	2	3	4	5
36. Is outgoing, sociable	1	2	3	4	5	1	2	3	4	5

	You						Dog				
	Disagree strongly	Disagree a little	Neither agree nor disagree	Agree a little	Agree strongly		Disagree strongly	Disagree a little	Neither agree nor disagree	Agree a little	Agree strongly
37. Is sensitive to the needs and feelings of others	1	2	3	4	5		1	2	3	4	5
38. Is planful, determined	1	2	3	4	5		1	2	3	4	5
39. Gets nervous easily	5	4	3	2	1		5	4	3	2	1
40. Appears to 'reflect', mull things over	1	2	3	4	5		1	2	3	4	5
41. Is co-operative	1	2	3	4	5		1	2	3	4	5
42. Is easily distracted	5	4	3	2	1		5	4	3	2	1
43. Is sophisticated	1	2	3	4	5		1	2	3	4	5

ANSWER

Use the following key to work out your respective totals, then copy them into the grid below to compare your personality to that of your dog.

- To calculate **Extraversion** (or its dog analogue, **Energy**), add up the circled responses for 1, 6, 11, 16, 21, 26, 31 and 36.
- To calculate **Agreeableness** (or its dog analogue, **Affection**), add up the circled responses for 2, 7, 12, 17, 22, 27, 32, 37 and 41.
- To calculate **Neuroticism** (or its dog analogue, **Emotional Reactivity**), add up the circled responses for 4, 9, 14, 19, 24, 29, 34 and 39.
- To calculate **Openness to Experience** (or its dog analogue, **Intelligence and Competence**), add up the circled responses for 5, 10, 15, 20, 25, 30, 35, 40 and 43.
- To calculate **Conscientiousness**, add up the circled responses for 3, 8, 13, 18, 23, 28, 33, 38, 42. This trait isn't really meaningful for dogs, but there's nothing to stop you calculating a **Conscientiousness** score for your chosen dog if you want to.

Copy your totals into the boxes below to compare yourself against your dog.

	You	Dog
Extraversion/Energy		
Agreeableness/Affection		
Neuroticism/Emotional Reactivity		
Openness to Experience / Intelligence and Competence		
Conscientiousness		

So what did you find? Is your dog a Bark Obama or J. K. Growling (intelligent, competent and conscientious), or more of a Katy Pawry or L. L. Drool J (extraverted, energetic, agreeable and affectionate)? Or does (s)he have the emotional reactivity of a Sinead O' Collar or William Shakespaw?

And, more importantly, how do you compare both to your dog and to your friends and family? If you would like to find out, but found this test a bit of a faff, don't worry; there's a much quicker and simpler personality test that you can use instead: it's time to finally find out . . .

The Truth about Cats and Dogs

In fact, this personality test has only one question:
Are you . . .

(a) a cat person?
(b) a dog person?
(c) both?
(d) neither?

ANSWER

The truth about cats and dogs – well, cat people and dog people – comes courtesy of our favourite anatine animal assessor, Sam Gosling, who gave visitors to his web site both a standard personality test (like the one in the previous section) and the cat/dog question. And here it comes . . .

Dog people score higher than cat people on **Extraversion**, **Agreeableness** and **Conscientiousness** (look back to the previous section for an explanation of these traits), while cat people score higher than dog people for **Neuroticism** and **Openness to Experience**. The 'boths' and 'neithers' tend to be somewhere in the middle, except that – like dog people – they score low for **Neuroticism**. The relationship between being a cat/dog person and these various personality traits did not differ by gender, although women are far more likely than men to be a cat person.

Now that we've found out the *real* truth about cats and dogs, just one question remains: what is it called when a cat wins a dog show?

A cat-has-trophy.

Dog-Person or Person-Dog?

Let's explore another 'real truth' about dogs, by comparing their behaviour with your own. Have a look at the pairs of pictures below.

1a

1b

2a

2b

Which picture in each pair do you prefer? And which picture, out of all four, did you find yourself looking at the longest?

ANSWER

Most people prefer the pictures on the left – the dogs or people greeting one another – to the pictures on the right – the dogs or people facing away from one another. This preference starts young; even six-month-old (human!) children spend longer looking at videos in which two people are facing one another than facing away. The reason is that we are an inherently social species. Humans are programmed from birth to crave and seek out social interaction, because this is how we learn everything from language to codes of behaviour.

Dogs are another inherently social species, and might therefore be expected to show the same pattern. And so it turns out. When dogs are shown images such as those on the previous page, they also look for longer at the social (left) than non-social pictures (right), whether the individuals greeting one another are dogs (top row) or humans (bottom row).

But there is one interesting difference between the two species: overall, most people look longest at the interacting dogs (top left), while most dogs look longest at the interacting people (bottom left). The researchers who conducted the study – Heini Törnqvist, Miiamaaria Kujala and their colleagues at the Universities of Helsinki and Aalto in Finland – suggest that this may be because it is more difficult to figure out what is going on in a social interaction when that interaction is between two members of a species other than one's own. An alternative possibility is that people prefer looking at dogs (i.e., the average person is a dog-person) while dogs prefer looking at people (i.e., the average dog is a person-dog). Actually, though, this second possibility is unlikely, as an earlier study by the same Finnish research group found that, when shown individual faces, dogs look longer at other dogs than at humans. This suggests that, although many humans will readily admit preferring dogs to people, dogs tend to prefer their own species.

This is not to say that dogs are uninterested in humans: quite the opposite. As any dog owner will tell you, dogs absolutely *love* both observing human–human interactions (as the study outlined above confirms) and – even better – interacting socially with humans themselves. The reason that dogs, more than any other species, enjoy human interaction is that we have selectively bred them to do so. Other *Canis* species – wolves, jackals, coyotes, wild dogs and dingoes – have no particular interest in our own. But for at least the past 15,000 years we have been selecting for breeding those individuals with the greatest affinity for humans, with the result that today's domestic dog really is man's best friend. As a consequence, dogs are capable of many human-like feats that are way beyond almost all other species, such as . . .

A Walking Dogtionary

As you'll know if you have one yourself, even the dimmest dogs are able to learn a handful of words, such as *walkies*, their own names and – in most cases – a few commands like *paw* and *roll over*. The brightest are able to learn hundreds of words. Indeed, the world's leading 'walking dogtionary', a female border collie named Chaser, has learned more than a thousand.

Two psychologists from South Carolina, John Pilley and Alliston Reid, got hold of Chaser after reading about Rico, a nine-year-old border collie who had learned the names of more than 200 objects. Although Rico's memory feat was impressive, Pilley and Reid wondered whether, by starting young – Chaser was just eight weeks old when she began her training – they could go one better. Or rather, 800 better. Over a period of around three years Chaser was taught one or two new words each day, which involved simply presenting her with the relevant object, and repeating its name between twenty and forty times. Chaser also underwent daily rehearsal testing, which involved retrieving a named toy from a set that usually contained around fifty items.

So, can you beat Chaser in a word-learning challenge? Opposite are ten of the objects whose labels were learned by this canny canine, along with the relevant word in Russian. (I chose a language with very different sound patterns from English, in order to try to make the task as hard for you as it would have been for Chaser.) Try to learn as many as possible, then test yourself later the same day (or, if you really want a challenge, the next day). Remember, Chaser learned two new words per day, so this is your target if you want to beat her.

Remember to come back later today (or tomorrow) to take the test on page 72.

malchik	medved	zmeya	sobaka	myach
chasy	noga	poyezd	loshad	ryba

A Walking Dogtionary: Test

In order to ensure a level playing field, I'll give you the same kind of test that Chaser had. Rather than giving you a picture and asking for the word (which, of course, you can't do with a dog!), I'll just give you each word and ask you to pick the relevant object. So, in the grid below, please find the *poyezd, noga, medved, malchik, sobaka, ryba, loshad, chasy, myach* and *zmeya*. Write the name of each in the relevant box, then look back to the previous page to check your answers.

Now, here's another test. From the grid below, please find the *yabloko.*

ANSWER

Did you manage to figure out which is the *yabloko*? You should have reasoned that the new word most likely goes with the object that you had not previously been taught a name for. That's right! The *yabloko* is the apple. Although this type of *exclusion learning* has often been claimed to be a uniquely human ability, Chaser passed this test with flying colours (though see *Something To Crow About?*). If you got more than two in the main memory test, then congratulations, you managed to avoid being outsmarted by a dog (though we can't really say for sure that you *beat* her, as Chaser was only given two new words per day; for all we know, she could have learned even more).

Just why are dogs so damn clever? According to the researchers behind the Rico study, the reason is that we humans have selectively bred dogs not just for their interest in humans (see previous section) but also for their ability – and desire – to understand and follow human commands. This is especially true for border collies, which, as sheep-herding dogs, were specifically bred for their intelligence and obedience.

Dogs' intelligence is another example of evolution in action (see *How the Giraffe Got His Neck*). Here we have an example of one species, the common ancestor of today's dogs and wolves, gradually evolving into another, as a result of the fact that individuals with a particular trait – in this case, understanding humans – enjoyed better breeding prospects than those without. This example of *artificial selection* (i.e., selection by human dog breeders), as opposed to the *natural selection* that happens in the wild, shows that we can not only *see* evolution happening before our very eyes but even *make* it happen. (See Web Links to read about a fascinating experiment in which scientists successfully replicated this domestication process on fast-forward with silver foxes, a species that is closely related to dogs and wolves but which had never been previously domesticated.)

Returning to Chaser, this canny collie was still able to pass the exclusion learning test, when re-tested *two years later*. So – don't look now – but opposite is one final test. If you really want to prove you're cleverer than a canine, put this book away, set a reminder on your calendar, and I'll see you in two years' time.

Web Links
There are plenty of videos of Chaser online. Here are just two:

https://www.youtube.com/watch?v=J982KYWohT8
https://www.youtube.com/watch?v=_6479QAJuz8

The following article discusses the *farm-fox experiment*, in which scientists attempted to recreate the process of domestication on a very short timescale:

http://www.americanscientist.org/issues/pub/early-canid-domestication-the-farm-fox-experiment

A Walking Dogtionary:
Two Years Later

Welcome back, and congratulations for surviving two more years in this uncertain world of ours. You don't look a day older. Now, from the grid below, please find the *lenta*. So you're not tempted to cheat, I won't give you the answer, but leave it to you – after you've made your selection, of course – to look back at the previous few pages for a reminder of how to work it out.

Who Man Being?*

We saw in the last section how modern dogs' ability to understand humans is an example of Darwinian evolution, albeit in this case by artificial – rather than natural – selection. Given that the assumption of Darwinian evolution is key to the central argument of this book – that the difference between humans and other animals is 'one of degree and not of kind' – I hope you will not mind if we set aside all the fun and games for a moment, and spend a chapter exploring the evidence for the theory of evolution, and what it means for Homo sapiens' place in the world.

Darwin's goal in *The Origin of Species*† was – you guessed it – to explain where different species come from. But what exactly is a species? Two animals belong to the same species if a male and female can successfully mate with each other, and any offspring can, in principle, have offspring of their own (i.e., they are fertile, rather than sterile). For example, the reason that horses and donkeys are classified as separate species is that – as we saw in **A Stable Personality?** – although they occasionally mate, the resulting mule or hinny is always sterile. In contrast, 'dog' is a single species because a male and female of very different breeds can produce a mongrel that can produce offspring of its own (provided that, despite its mongrel looks, it can still find someone to love it).

If you want to find out whether two animals are generally considered to be of the same species, all you need to do is look up their proper scientific names. On the whole, I've spared you these, in favour of terms like 'horse' and 'donkey' (informal labels that – in fact – don't always map neatly on to a particular species). But every

---

* The section was inspired by Yuval Noah Harari's best-seller *Sapiens: A Brief History of Humankind*.

† Or, to give it its full title, *On the Origin of Species by Means of Natural Selection, or the Preservation of Favoured Races in the Struggle for Life*.

animal has a scientific name consisting of its *genus*, followed by its *species*. A genus is group of species that share a common ancestor. For example, the genus *Equus* consists of seven species: *Equus africanus* (African Wild Ass), *Equus ferus* (Wild Horse), *Equus grevyi* (Grévy's Zebra), *Equus hemionus* (Onager), *Equus kiang* (Kiang), *Equus quagga* (Plains Zebra) and *Equus zebra* (Mountain Zebra). So notice that, while we might talk informally about 'zebras', there are actually three zebra species that are entirely separate (i.e., they cannot successfully mate, yielding fertile offspring).

On the other hand, although a species may be divided into sub-species, these distinctions are not important from a reproductive standpoint: i.e., individuals from different sub-species *can* successfully mate to produce fertile offspring. For example, the species *Equus ferus* has three known sub-species: *Equus ferus caballus* (your bog-standard horse), *Equus ferus przewalskii* (an endangered Mongolian wild horse) and *Equus ferus ferus* (the tarpan, which became extinct in 1909). But all are members of the same species: a *caballus* and a *przewalskii* could get together, form a stable relationship (sorry!) and produce fertile offspring. This isn't just a theoretical possibility: in the mid-1970s a *przewalskii* stallion named Csar mated with three (sadly anonymous) Welsh ponies at Regent's Park Zoo, and two of the three offspring – one colt, one filly – went on to have offspring of their own.*

So, now we've established what we mean by a species, we can play a guessing game. For each genus shown overleaf, how many different species survive today?

---

* Under the old rules this would be enough evidence to say that *caballus* and *przewalskii* are – by definition – members of the same species. In the modern era, however, we can also use DNA sequencing to draw the boundaries; and occasionally – as in this case – there's room for debate about whether or not the two methods give the same answer.

Felis (small-medium size cat, including the domestic cat)	
Phascolarctos (e.g., koala)	
Ursus (bears)	
Giraffa (you can guess this one!)	
Loxodonta (African elephants)	
Elephas (Asian elephants)	
Canis (wolves and dogs, including domestic dogs)	
Homo (man)	

ANSWER

Although generally there are no more than a handful of species remaining in each genus (and sometimes only one), most had many more species, which are now extinct.*

But here's the punchline: our own genus is no exception. Although only *Homo sapiens* survives today, there have existed at least six other species that share a common ancestor: *Homo erectus*, *Homo floresiensis*, *Homo habilis*, *Homo heidelbergensis*, *Homo naledi* and *Homo neanderthalensis*. As I have stressed throughout this book, there is really nothing special about *Homo sapiens*. In fact, despite our iPads and penicillin, we're not even the most successful human species, at least in terms of survival. *Homo erectus* (upright man, who lived in East Asia) survived for almost two million years: ten times as long as we have managed so far.

How do we know all this? Well, up until fairly recently all we had was the fossil record. But since the advent of DNA sequencing we have been able to trace our ancestry with considerable precision. We now know that modern humans (*Homo sapiens*) and Neanderthals (*Homo neanderthalensis*) are not only closely related but actually co-existed and even – very occasionally – copulated. So what happened to the Neanderthals? Until a few years ago, many experts thought that this interbreeding was sufficient to merge

Homo (man)	1
Canis (wolves and dogs, including domestic dogs)	7–10
Elephas (Asian elephants)	1
Loxodonta (African elephants)	2
Giraffa (you can guess this one!)	1
Ursus (bears)	4
Phascolarctos (e.g., koala)	1
Felis (small-medium size cat, including the domestic cat)	6

* In fact, of all the species that have ever existed on earth (including both plants and animals), more than 99 per cent are now extinct.

the two species. However, analyses published in 2011 and 2012 suggest that the amount of Neanderthal DNA in modern-day humans is too low for this to be plausible. In all likelihood, the brutal truth is that sapiens simply wiped out the neanderthals (in Eurasia), as well as the erectus (in East Asia). This is the familiar *Out of Africa* theory: that all modern-day humans are descended from a population of sapiens in East Africa, who began to migrate perhaps around 70,000 years ago.

If you're wondering what all this has got to do with evolution, the answer is that evidence collected using the same technique, DNA sequencing, clearly shows that *Homo sapiens* and other human species (and more distantly, chimpanzees and bonobos; see Are You Smarter than a Chimpanzee? #1) share a common ancestor. The only other possibility is that a creator is continually making new species from scratch but – for reasons best known to him/her/itself – is embedding in each a genetic code designed to trick us into believing in evolution. There is nothing preventing a creationist from making this argument, of course, but even if we accept it, there is still an important detail to iron out. When God made man in his image, what species was this: *Homo sapiens*, one of the others, or even more than one? If, as they might well have done, neanderthals had survived too, modern religions would presumably have looked rather different.

At the risk of using a sledgehammer to crack a nut, let me end this section with some particularly clear evidence not just for evolution but for Darwinian evolution by natural selection. Many sceptics will say something along the lines of 'Fossils, DNA? Whatever. I'm not believing anything that I can't see with my own eyes.' In fact, you *can* see evolution by natural selection happening with your own eyes, and over a relatively short time span.* In 1995 a team of scientists introduced the Cuban brown anole lizard (*Anolis sagrei*, since we're doing this now) into some islands in Florida that were previously home to just one other Anolis species (*Anolis*

---

* Another example is an all-too-real threat to our species' current position as top dog: the evolution of bacteria to resist antibiotics.

carolinensis). Over time the native species moved higher and higher into the trees and – crucially – evolved to have larger toe-pads, covered with a greater number of sticky scales. And all this took just twenty generations (spanning fifteen years).

So the evidence that *Homo sapiens* is – like all species – a product of evolution is overwhelming. But why, in that case, did this one species come to dominate not just all other species of humans, but all other species full stop? We can only speculate. But, as we will see later (in Why Can't We Talk to the Animals?), perhaps the most plausible theory is that the hand-in-hand development of language and co-operation allowed *Homo sapiens* to construct schemes – and weapons – that were ruthlessly effective.

Web Links
https://en.wikipedia.org/wiki/Felis
https://en.wikipedia.org/wiki/Canis
https://en.wikipedia.org/wiki/Ursus
https://en.wikipedia.org/wiki/Phascolarctos
https://en.wikipedia.org/wiki/African_elephant

A Formidable Problem

Although true co-operation is arguably unique to humans, even some humble insects show a kind of unthinking quasi-co-operation which allows them to perform some very sophisticated tasks, such as finding the most efficient route to a destination.

But how do you compare? Suppose you are a travelling salesman living in Ashton-in-Makerfield (a former coalmining town in Greater Manchester). On one particular day, you need to make sales calls in Bolton, Chorlton-cum-Hardy, Dunham Massey and East Warrington, then return home to Ashton. The map below shows the travelling time in minutes between each of these places (for simplicity's sake, assume that these times are fixed and do not change depending on the traffic). What is the best order in which to visit these places to minimise your time spent on the road? (NB: you are not allowed to visit any of these places more than once, even if it would shorten your travelling time.)

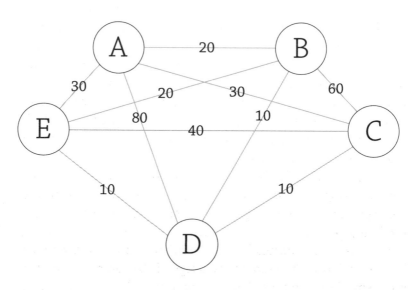

ANSWER

The best possible solution for this particular problem has a total journey time of 90 minutes. Did you get anywhere close to that? If not, here's a strategy that – at least for small-scale problems such as this one – usually gives you a pretty good solution (though not necessarily the best one possible): the 'nearest neighbour' method. Every time you set off on the next leg of your journey, simply choose the place with the shortest travelling time, excluding any that you have already visited (except, of course, your home town on the final leg). This would give you either A, B, D, C, E, A or A, B, D, E, C, A (depending on which of the two 10-minute journeys you picked when setting off from D), both of which come in at 110 minutes (20+10+10+40+30). Now that you have a pretty good route, see if you can shave off 20 minutes by trial and error.

Have you got it yet? Here's one final hint: if you start off with the latter of the two possible 'nearest neighbour' routes (A, B, D, E, C, A), you can find the shortest route simply by swapping two of your stops.

That's right! Swapping D and E gives you A, B, E, D, C, A, for a total travelling time of 90 minutes (20+20+10+10+30).

Despite over fifty years' work on the travelling salesman problem, mathematicians have not been able to come up with a solution that is guaranteed to yield the best possible route, other than the 'brute force' method of simply comparing every possible route, which quickly becomes impossible if more cities are added (with just fifteen cities, there are 87 billion possible routes). In fact, most experts think that an algorithm that gives the best route every time is not possible.

In their search for better approximate solutions, scientists are increasingly taking their inspiration from the animal world. One approach is based on genetics. First, a number of routes are randomly generated. Most of these won't be much good at all, but the best (i.e., shortest) routes are then 'mated' with random 'mutations' (usually removing two legs of the route then replacing them). After a hundred or so generations, this process usually yields a pretty good route, even for much larger problems with, say, twenty-five cities.

An alternative approach is based on ants. Ants are fascinating in that they have a kind of collective (or 'swarm') intelligence.* Rather like individual cells in the brain, each individual ant can perform only very simple tasks but can combine with other ants to show complex behaviour. Consequently, in many ways, it is the colony – rather than the individual ant – that functions as an organism. No individual ant 'knows' that a nest needs building or food gathering; this 'knowledge' exists only at the level of the colony as a whole (many researchers explain ants' hunting and feeding behaviour in terms of a 'common stomach').

Perhaps the most famous example of this collective intelligence is how ants find the shortest route from the nest to a particular food source. No individual ant works out the shortest route, but the colony as a whole is quickly able to do so. Ants explore the area around the nest at random, leaving chemical pheromone trails as they go. When one ant happens to find some food, it takes some back to the nest, following the pheromone trail that it left on its outward trip. When another ant comes across this now double-strength trail, it follows it to the food (either directly or via a futile trip back to the nest, depending on whether or not it happened to pick the right direction when following it). This ant then takes some food back to the nest, strengthening the pheromone trail still further. All that an individual ant 'knows' is to follow the strongest pheromone trail that he stumbles upon; but collectively, the colony 'knows' the best route to the food.

This gave Italian Artificial-Intelligence researcher Marco Dorigo an idea: an ant colony – or, at least, a computer simulation of one – might be able to come up with better solutions to the travelling salesman problem. Each virtual ant is placed in a random starting

---

* This isn't to say that all ants contribute to the work of the swarm. In fact, one recent study found that about a quarter are slackers who never get their hands dirty at all: http://www.sciencemag.org/news/2015/10/most-worker-ants-are-slackers

city, then walks to a not-yet-visited city chosen at random (or, if none remains, back to his starting city). These virtual ants leave virtual pheromone trails as they go. The clever part is that, unlike real ants, these virtual ants have a fixed amount of pheromone that they distribute evenly over the entire 'tour'. This means that ants who followed a shorter route will leave a stronger trail than those who followed a longer route and who, consequently, had to spread their pheromone allocation more thinly. Each ant then starts out again, but this time following the strongest pheromone trail that he comes across. After a few hundred runs through this procedure, these virtual ants will have discovered a route that is, while not always the best possible, right up there with the best achieved by the mathematicians.

There's only one word for that: *brilli-ant* (or, as the French would say, *formi-dable*).

Web Links
http://www.wired.com/2013/01/traveling-salesman-problem/
The map in this section is based on the example at: http://algoviz.org/OpenDSA/dev/OpenDSA/Modules/NPComplete.odsa

The Tower of Han(t)oi

The **Formidable** (travelling salesman) **Problem** of the previous section was solved not by real ants but by a computer algorithm inspired by their behaviour. If this has left you with ants in your pants, or feeling at bit antsy, don't worry. It's now time to up the ante (OK, I'll stop now) by pitting you against real live ants: specifically – if this isn't an oxymoron – Australian Argentine ants. The task at which you and the ants will be competing is a famous problem known as the Tower (or Towers) of Hanoi. Your task is to move the tower, which consists of three discs, from the leftmost to the rightmost of three posts.

The game has three rules: (1) you may only move one disc at a time; (2) you may only ever play a smaller disc on top of a larger disc (never vice versa); (3) make as few moves as possible.

Because it's difficult to work through the problem in your head, you may like to cut out the discs below (or download the template from www.areyousmarterthanachimpanzee.com).

Are you ready for the answer yet? If you need a hint, the minimum possible number of moves is seven.

PS: This problem is based on the Indian legend of the Tower of Brahma, in which the discs are made of gold. Oh, and there are sixty-four of them. Doesn't sound *that* hard, right? Wrong! At one move per second, the optimal solution would take almost 600 billion years to carry out. Good luck with that!

ANSWER

Right, here's the solution:

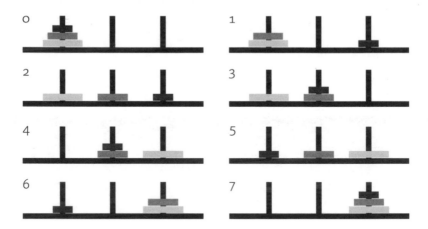

It's not that difficult once you get the hang of it, right? In fact, there is a general solution that works for any number of discs:

- For an odd number of discs, make the only possible legal move (which will sometimes be rightward, sometimes leftward) between (1) pegs A and C, then (2) pegs A and B, then (3) pegs C and B. Repeat these three steps until complete.
- For an even number of discs, make the only possible legal move (which will sometimes be rightward, sometimes leftward) between (1) pegs A and B, then (2) pegs A and C, then (3) pegs B and C. Repeat these three steps until complete.

At this point you are probably wondering how you would go about explaining the game to a colony of ants (not to mention whether or not they could carry the discs). The answer is: you can't. In order to get ants to solve the problem, you have to translate it into something they can understand: finding food. This involves turning the game into a maze, which sounds complicated but is actually very simple. All you need is a maze in which, at every junction, the possible paths that an ant can take correspond to the possible moves in the game.

NEST

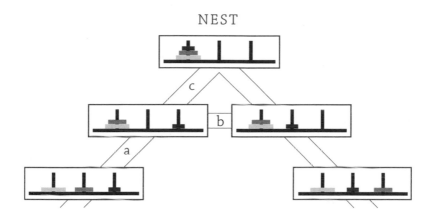

For example, the first junction that the ant comes to after leaving the nest has two branches, corresponding to the two possible first moves. The next junction has three branches, which correspond to three moves now available. For example, if the ant chose the left branch (dark grey disc to third peg) at the first junction, the three branches now correspond to: (a) light grey disc to middle peg (the best move); (b) dark grey disc to middle peg; and (c) dark grey disc back to first peg (a legal, if clearly stupid, move, which – in the ant version – entails heading straight back to the nest). In this way the entire Tower of Hanoi game can be mapped out as a rather beautiful maze of hexagons forming two mirror-image triangles, (see web link).

Perhaps unsurprisingly, given what we learned about ants' food-finding abilities in the previous section, over 90 per cent of the ant colonies tested were able to find the shortest route through the maze (corresponding to the seven-move solution shown above) in less than an hour. Even more impressively, when the researchers blocked off some of the paths, the colonies were able to find the best possible remaining route of those still available. This challenges the conventional wisdom that ant colonies are unable to respond to rapid changes in the environment.

Before the study, most experts would have assumed that the ants would continue to follow the original – though now useless – trail.

Of course, no individual ant would be able to solve the Tower of Hanoi problem on its own. The solution resides not in the brain of the individual ant but in the 'brain' of the colony as a whole. But it's not only ants who show this *swarm intelligence*; humans do too . . .

Web Link
See the full maze at: http://jeb.biologists.org/content/214/1/50(fig36)

Losing Your Marbles

First, let's see how clever you are when you have to work something out on your own. How many stones do you think are in this jar?

You don't have a clue, do you? Neither, individually, do any of the readers of this book. Collectively, however, you do know. If you all make a (sensible!) guess and we take the average, the estimate that you collectively come up with will be almost exactly spot on. But don't just take my word for it. Go to http://unanimous.ai/rocks and enter your guess. The web site will then tell you the current average (including your guess) and how far away this is from the actual number. Typically, the average aggregate guess is only around 1–2 per cent out. (The study is traditionally run with a jar of marbles, hence the title.)

How come? Well, assuming you can all count and have learned some basic maths, everyone's guess will be somewhere in the right ballpark. As a result, every guess that is – say – 50 too high will be balanced out by a guess that is 50 too low; every guess that is – say – 100 too high will be balanced out by a guess that is 100 too low and so on.

The first demonstration of this phenomenon was carried out by Francis Galton (see Are You Big-Headed?). Galton asked 800 visitors to a 1907 livestock exhibition to guess the weight of an ox, and found that the 'middlemost estimate' was just 0.8 per cent out.* In the modern era, companies including Google, Microsoft and HP have tried to harness this 'wisdom of the crowd' to predict everything from elections to share prices and sports results. One commonly used technique is the prediction market, in which people are asked to place bets on particular outcomes. Usually, researchers give participants a small budget and then pay out winnings in real money, giving participants an incentive to make their predictions as accurate as possible.

What about animals? We have already seen how ants' swarm intelligence can help with navigation (see A Formidable Problem). Another problem that ants (and bees) solve using swarm intelligence is deciding where to build a nest. But not everyone in the colony gets a vote. A small number of scouts explore potential locations, with each 'voting' for a particular one simply by hanging around there. Once the number of scouts at a particular location reaches a particular threshold (or 'quorum'), the vote is over. The scouts return to the group and direct them to the winning location (in the case of bees, by using the famous 'waggle dance'). The

---

* In his original report Galton used the median, which you calculate by lining up all the guesses in order and choosing the one in the middle. However, if you use the mean (i.e., add up all of the guesses and divide by 800) then the crowd's guess (1197 pounds) is essentially perfect (the ox weighed 1198 pounds).

reason that this counts as an example of swarm *intelligence*, rather than merely mindless groupthink, is that this process invariably results in the selection of the best possible site from all of the alternatives available.

Fish face a similar problem of choosing a place to live that has all the required amenities (food, oxygen and sunlight). As for humans, a key factor in the decision is the local schools (pun very much intended). Fish form into large swarms, called 'schools', which allow them to make collective, and therefore better, decisions. If an individual fish is deciding where to hang out, it performs what's called a 'random walk' (though, of course, it's actually a swim). This involves wandering about at random but changing direction more quickly when the environment is getting worse than when it is getting better. Because any individual fish can only explore quite a small area, this process isn't very effective. The solution is to join a school. This involves getting close to other fish, but not too close (which would reduce the overall area covered) and swimming in the same direction. This ensures that when a handful of fish stumble across a prime location, their schoolmates rapidly join them.

Swarm intelligence is yet another example of how unintelligent design by means of natural selection (Richard Dawkins' 'Blind Watchmaker') can hit upon ingenious solutions. No designer – neither an individual fish genius nor nature itself – came up with the idea. All that is needed are random gene mutations that cause fish to want to get close to and follow one another. Because the result is a school, which makes better decisions than any individual, fish with these mutations thrive at the expense of their rivals, and the cycle continues. Just like the random walk of the school itself, the random walk of evolution uncovers a solution that is too brilliant to have been dreamed up by any individual.

Counting Cheep

In the previous section (Losing Your Marbles) we looked at your ability to roughly reckon large quantities without counting. Here we're going to do something similar, but for much smaller quantities. In each of the pairs below, either the right- or left-hand box has more dots. Your job is simply to guess – as quickly as possible, and without counting – which box in each pair has more dots.

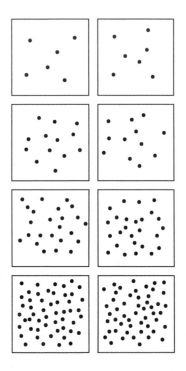

ANSWER

OK, here are the numbers of dots in each box:

(a) 6 vs 7; (b) 13 vs 12; (c) 24 vs 26; (d) 52 vs 51.

These pairs range from the easy to the tricky, to the very difficult, to the frankly impossible as the overall number of dots increases, and the difference between the two, in proportional terms, decreases.

This was a test of your 'analogue magnitude system'. Most experts believe that animals, including humans, have two different systems for processing numbers of objects. The 'object-file system' involves keeping in memory a separate file for each object (in much the same way as you might keep files on your computer desktop). Because these files are very detailed – including information about the colour, size, shape and motion of each object – you can keep, at most, four files in memory at any one time. The plus side is that this system is not only highly accurate but also able to do calculations without even thinking about it. Suppose you are keeping an eye on your three children while they play in the garden. If you want to know whether one of them has run off, you don't have to – as Americans say – do the math ('I counted 3 before, but now I only count 2. Since 3-2=1, then 1 must have gone missing'); your object-file system did it for you the instant that you glanced out of the window. Your analogue magnitude system, on the other hand, can't do any math(s) at all: all it is good for is estimating the rough number of things in a group.

Now, everyone agrees that we use the analogue magnitude system whenever there are more than four things to keep track of, as in the test on the last page (we have no choice, as the object-file system can only count up to four). A more controversial question is whether we *always* use the object-file system for quantities below four, or whether – if we can get away with it – we sometimes use the rough-and-ready analogue magnitude system instead.

In order to find out, a group of Italian scientists decided to test – cuteness alert! – new-born baby chicks. The chicks had to learn that food was always hidden behind (for example) a three-dot pattern, but never a two-dot pattern. By varying the numbers of dots, the experimenters were able to test chicks on various comparisons, just like I did with you on the last page. The findings were remarkable: these chicks – new-born remember – passed the test for 2 vs 3; 2 vs 8; 6 vs 9; 8 vs 14; 4 vs 6; and 4 vs 8 dots.

'Baby chicks can count' screamed the headlines (just Google it), but they had it exactly backwards: the whole point is that the chicks don't *have* to count, and for the larger numbers absolutely *cannot* be counting. Rather they are using their analogue magnitude system to get a *rough idea* of the number of dots (and therefore whether the picture with *about that many* is the one that hides the food).

Did the chicks ever 'count'? In other words, thinking back to our original question, did they use the more precise object-file system when the opportunity arose (i.e., whenever there were four dots or fewer), or did they just stick to the analogue magnitude system throughout? In fact, the evidence seems to favour the latter. Particularly key here are the 2 vs 8 trials. We know that they must have been using the analogue magnitude system for the eight-dot pattern (as the object-file system can only count up to four). So, unless they were using different systems for the different patterns – which presumably would make comparing the two rather difficult – the chicks were presumably sticking to the analogue magnitude system.*

The researchers conclude that, even if other animals have some kind of 'number sense . . . abstract mathematical logic remains a prerogative of our species'. But are they right, or could it be that some animals can in fact do maths, and even reason logically? We will find out later in the sections **Mathemalex** and **Something To Crow About?**. In the meantime, let's give three (count 'em) object-file cheers for these champion chicks: Cheep, Cheep . . . Hooray!

---

* This is not to say that they *never* count. One possibility discussed by the researchers is that quantities of four and below trigger the analogue magnitude system when presented as a group, but the object-file system when presented one after another.

Circle of Life

In fact, these baby chicks are so clever, they are able to show human-like performance in one particular set-up in which their fellow birds (pigeons) and even primates (baboons) do not. But what is human-like performance on this task? Let's find out by testing a human: you. Which is bigger: the grey circle on the left or the one on the right?

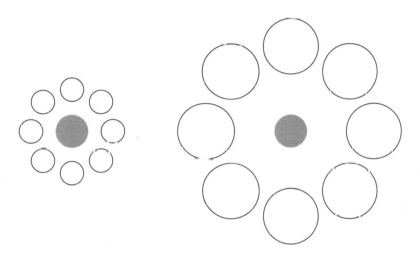

ANSWER

Even if you suspected a trick and said that the two circles were the same size (which they are), it's impossible to convince yourself, isn't it? Over a hundred years after its discovery by German psychologist Hermann Ebbinghaus (1850–1909), the causes of this *Ebbinghaus illusion* remain poorly understood. The intuitive explanation that you have probably already come up with yourself (i.e., the grey circle looks big in comparison with the small ones and small in comparison with the big ones) is about as far as we have got. One thing is clear: we seem to 'learn' to see the illusion. Seven-year-old children are misled to a far lesser extent than adults. Yes, in one specific sense we adults actually see the world less accurately than we did as children.*

Given that, for humans, experiencing the illusion seems to be a sign of being *more* advanced (or, at least, older), it is surprising that these baby chicks experience the illusion, while more 'advanced' animals (adult pigeons and baboons) do not. How do we know that chicks see the illusion? Well, first each chick was taught – using two circles that *really were* of different sizes – that food can be found hidden behind (say) the larger circle of the pair. The chick was then shown the pair of circles that you saw on the last page, and showed susceptibility to the illusion by looking behind the (apparently) 'larger' circle. And, remember, unlike humans, they didn't need to 'learn' to see the illusion at all: they were just four days old (aw!).

The authors of the chick study point out that we should be cautious when comparing the chicks with pigeons, as the pigeons were faced with a more difficult, and less natural, task (they had to peck keys, rather than just look for food). But if the difference holds after ruling out this possibility, this suggests some-

---

* Interestingly, members of a remote culture with no words for shapes (the Himba) are also less susceptible to the illusion. However, this seems to be more because they place greater importance on examining local details than because of their lack of shape names per se.

thing particularly intriguing: that humans and domestic chickens independently evolved the brain mechanisms that cause the illusion, while species more closely related to each (i.e., baboons and pigeons) did not.

In fact, surprising though it may seem, chickens' vision is actually unusually impressive in many different ways, as we will see later in the section For Eagle-Eyed Readers.

Pige-lusi-on #1

Sticking with the subject of birds and illusions, we move from a visual illusion to a psychological illusion: superstition. Superstition is the illusion that we can temporarily bend the rules of the universe in our favour by, for example, touching a piece of wood or muttering a magical phrase. When it is described in such blunt terms, it is clear that superstition is utterly irrational. Yet many otherwise rational people will go out of their way to avoid walking under a ladder, or go weak at the knees if somebody opens an umbrella indoors. How about you? Do you agree with Stevie Wonder that superstition ain't the way, or are you as superstitious as a . . .

. . . pigeon? That's right, our feathered friends are among the most superstitious creatures on the planet. But are you? Fill in the short questionnaire below to find out.

	Definitely NO ⟷ Definitely YES				
Have you avoided walking under a ladder because it is associated with bad luck?	1	2	3	4	5
Would you be anxious about breaking a mirror because it is thought to cause bad luck?	1	2	3	4	5
Are you superstitious about the number 13?	1	2	3	4	5
Do you say 'fingers crossed' or actually cross your fingers?	1	2	3	4	5
Do you say 'touch wood' or actually touch or knock on wood	1	2	3	4	5
Do you sometimes carry a lucky charm or object?	1	2	3	4	5

Add up your scores to find your superstition total. The average total, when this test was given to almost 5,000 internet responders, was around 14/30 for men and 18/30 for women. Yes, it's true: for whatever reason, women are – on average – more superstitious than men, with the difference particularly pronounced on the last three questions (which involve *actively doing* something, rather than just *avoiding* something). So if you scored more than a couple of points higher or lower than these averages, then you are officially more or less superstitious than average for your gender. This study also found a link with neuroticism: people with high scores tend to be more neurotic, as measured by a standard personality-test measure (see A Stable Personality?). Neurotic people give particularly high scores on the first three questions. That is, as one might expect, they're more worried about accidentally triggering bad luck than about taking steps to ensure good luck.

For me, what is surprising about these findings is just how high the average scores are. We all know that superstitions are useless, right? Yet the average person scores somewhere around the middle of the scale for each question. That is, the average person either does all of these things to a moderate extent or does some of them often and others never. But why does anyone do *any* of these things *ever*? Do we seriously think they will work?

If so, then we are no smarter than pigeons. In 1948, the psychologist B. F. Skinner published the findings of his studies on *Superstition in the Pigeon*.* Skinner set up a cage with a movable food hopper. Every fifteen seconds the hopper swung into place next to the cage, allowing the pigeon to feast for five seconds, before swinging away to begin the countdown again. Although the hopper appeared every fifteen seconds no matter what, the birds developed elaborate rituals to – as they thought – bring it back

---

* *The Simpsons'* Principal Skinner is named after B. F. Skinner, presumably because the two share a similarly mechanistic approach to education. In one *Simpsons* episode, Skinner even name-checks his namesake, saying, 'Every good scientist is half B. F. Skinner and half P. T. Barnum [the showman and circus proprietor].'

again. One span on the spot, one nodded, one repeatedly tossed its head into the upper corner of the cage, one swept the floor and two swung like pendulums. In fact, of eight birds tested, only two failed to develop one of these superstitions.

How did these superstitions develop? In each case, the bird just so happened to be performing this particular behaviour when the food arrived (and, in many cases, was still doing so when it next reappeared). Any remaining doubt in the pigeon's mind regarding the link between spinning/nodding/sweeping etc. and food delivery would have been extinguished when – lo and behold – doing so 'caused' the hopper to return. Of course, at any point a curious pigeon could have checked to see whether stopping the ritual would stop the food delivery, but why take the risk? Presumably something similar is going on with people who cross their fingers or touch wood. Sure, they know there's probably nothing to it, and they could stop with no ill effects, but why take the risk?

Pige-lusi-on #2

Still on the subject of pigeons and illusions, we come now to perhaps the most famous visual illusion of all time: the Müller-Lyer illusion. The two lines below are exactly equal in length (measure them if you don't believe me), but the second looks much longer, right?

Although the illusion is over a hundred years old, we still don't understand exactly what causes it. The standard textbook story is that when we see a configuration of the second type in the real world, it's usually in the context of – for example – a faraway wall meeting the floor, as shown here (imagine you are peering in at the window).

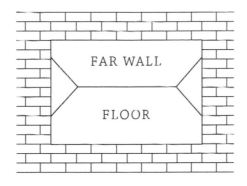

Because we know that faraway objects are actually larger than they appear, we mentally boost the length of this line. Evidence for this theory comes from the fact that members of remote foraging tribes, who never encounter any man-made buildings – and so don't spontaneously perceive these types of configurations as 'faraway-wall meets floor' – aren't susceptible to the illusion.

Although this finding suggests that this theory is probably along the right lines (sorry!), it doesn't tell the whole story.* Look what happens to the apparent length of the line if we make the brackets bigger:

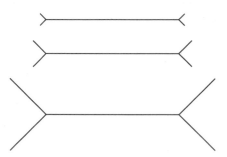

It seems to get shorter, right? The faraway-wall-meets-floor theory can't explain this pattern, as the angles are identical in all three versions, meaning that the 'wall' should be perceived as the same distance away, and get the same mental length-boost, in each case. Instead, what seems to be happening is some form of Ebbinghaus illusion, as discussed in Circle of Life surrounding something (here, the horizontal line) with big things (here, the brackets) makes it look smaller by comparison.

How can we test this possibility? Well, as we saw in Circle of Life, humans fall for the Ebbinghaus illusion, whereas pigeons (probably) don't. So if we get the same pattern with this business of making the Müller-Lyer brackets bigger – humans see a shorter line, while pigeons don't – this would support the idea that the Ebbinghaus illusion is the culprit.

As you can imagine, running this type of study with pigeons is a bit of a faff. First you have to train them to peck one key for long lines and another for short lines. Then you have to train them to ignore the brackets (by having brackets above and below the line, pointing in different directions). You're then ready for

---

* See the chapter *Lyer, Lyer* in my previous book *Psy-Q* for another fly-in-the-ointment for this theory.

the test session, which involves varying the size of the brackets – while holding constant the length of the horizontal line – and seeing how often the pigeons peck the 'long' key. Fortunately for us, some enterprising researchers at Japan's Kyoto University went through all of this rigmarole with three homing pigeons (the none-too-Japanese-sounding Kay, Indy and Hans).

So what happened? Remember that, for humans – as you saw for yourself – increasing the size of the brackets makes the horizontal line look shorter. For pigeons, increasing the size of the brackets makes the horizontal line look *longer* (i.e., they give more 'long' key pecks for a given horizontal line length). It's not clear why this is, but perhaps the bigger brackets somehow cause the pigeons to treat the Müller-Lyer figure as a single long, bendy line with some extra bits (unlike humans, pigeons cannot, of course, be asked to report the length of the horizontal line only).

Whether or not this is the case, the fact that pigeons didn't perceive a shortening of the line suggests that this tendency in humans is indeed some kind of Ebbinghaus effect. So why do humans perceive Ebbinghaus effects while pigeons don't? The researchers behind the pigeon study suggest that because pigeons eat grains – which are uniformly tiny – they are not generally in the business of comparing the size of different foods (or, for that matter, objects of any kind). Then again, baby chicks also eat grain but – as we saw earlier – do perceive the Ebbinghaus illusion.

In short, more research is needed (which is scientist-speak for 'we don't have a clue what's going on'). Wouldn't things be so much easier if the birds could just tell us what they can see?

Mathemalex

One bird that could do so, in theory, is the parrot. Humans aside, this famously talkative bird is arguably top of the class for English. And also, somewhat surprisingly, maths. If you think that doesn't quite add up, read on . . .

Alex, an African Grey parrot, is well known in linguistic circles for having the gift of the gab. In fact, by the time his owner, Irene Pepperberg, decided to investigate Alex's mathematical abilities, he had already learned the numbers 1 to 6, and could correctly answer 'how many' questions for collections of up to six items. He could even give the correct answer to questions such as 'How many blue blocks?' when shown a mixed collection of blue blocks, blue balls, green blocks and green balls.

So Alex certainly shows some impressive counting abilities, but adding up is a whole different ball game. Would Alex go one step further and pass a test of addition? This test involved showing him the number of nuts hidden under each of two separate cups, then asking, 'How many nut total?' Importantly, the nuts were hidden when the question was asked, meaning that Alex could not simply count. Rather, he had to remember the number of nuts under each cup, and somehow combine the two.*

Alex did pass the test, and with flying colours. Even for the hardest sums (those where the answer was 5 or 6), he scored 81 per cent, beating chance performance by a large margin. And he didn't just learn the answers by rote. Alex's performance was similar (84

---

* If we're determined to quibble, I suppose Alex could theoretically picture all six nuts as a single spread-out collection, then count them, or count the number under the first cup and then 'count on'. However, both of these strategies seem to be addition of some kind, and neither is necessarily more straightforward than whatever way of combining the two numbers would count as 'real' addition. So let's not quibble.

per cent correct) if we look only at his first attempt at each sum.

Interestingly, some of the sums at which Alex succeeded included zero (e.g., five nuts under one cup, zero under the other). Does this mean that Alex understands the concept of zero? If so, this would be news indeed, and suggest that his understanding extends beyond the realm of natural numbers – numbers that can be used to count things – and into the world of integers: positive and negative whole numbers, as well as zero. At least according to sum (sorry, *some*): mathematicians are divided as to whether or not zero is a natural number. Although it might seem obvious to us that one can have – for example – 'zero nuts', this very concept is something of a paradox: how can nothing be something? If you don't have anything, how is there any quantity of nuts for 'zero' to refer to (why is it zero *nuts* rather than – say – zero chickens)? One thing is for sure: zero is a relatively recent human invention, the first recorded use being around 3 BC, and many perfectly good counting systems – roman numerals, for example – managed without it. So if Alex has figured out the concept, we could make an argument that he's more mathematically sophisticated than pretty much anyone born before Jesus.

So has he? The early signs were promising. In one study Alex was shown – for example – one red block, two blue blocks and three green blocks, and asked, 'What colour two?' (i.e., 'What is the colour of the set of size two?'). Alex not only gave the correct response to such questions (here, 'blue') but spontaneously began to answer 'none' when appropriate (e.g., when asked 'What colour four' for the 1-, 2- and 3-set collection above). Alas, when – in the counting study – Pepperberg showed Alex two empty cups and asked 'How many nut', he was flummoxed.

Or was he? On over half of the 'zero' trials Alex looked at the cups and said . . . nothing.

Whether or not he understands the concept of zero, we can certainly say that Alex's mathematical skills lie outside parrots' normal abilities. But could a parrot go one stage further and show (cue spooky music) . . .

Parrotnormal Activity

Inspired by reports of Alex, New Yorker Aimée Morgana began training her own parrot, N'kisi, to speak. But when N'kisi began talking, Aimée got quite a shock: her parrot seemed to be psychic . . .

One day Aimée picked up the phone to call her friend Rob, and was just in the process of looking up the number, when N'kisi piped up with 'Hi, Rob'. On another occasion, Aimée was downstairs looking at a picture of a purple car, when N'kisi called down from upstairs, 'Look at the pretty purple'. Most spookily of all, on a couple of occasions, N'kisi – who generally slept by his owner's bed – woke her up by narrating her dreams (for example, saying 'See, that's a bottle', when Aimée was dreaming about holding a brown medicine bottle).

At this point, Aimée called in Rupert Sheldrake, a biologist who specialises in the paranormal, to investigate whether N'kisi's abilities would hold up under test conditions. Let's play along: do you have parrotnormal abilities . . .?

To find out, you will need a friend to play the part of Aimée, the 'sender'. Familiarise yourself with the ten pictures below (write a list if you like), then pass the book to your friend. On each trial your friend should pick one of the pictures and concentrate on it (ideally for two minutes, but less is fine if you're pushed for time). Your job is simply to guess which picture was being 'sent' on each of the ten trials.*

---

* Incidentally, I once took part in a similar study run by an undergraduate psychology student at my university (the University of Liverpool). After I completed the test (which did not suggest any psychic abilities), the student sat me down for a debrief, and explained that he was hoping to replicate a previous study that had indeed found above-chance performance on such a test. 'Right,' I said, 'so you were in the room next door, looking at the pictures and trying to send me the right answer?' 'No', he replied, looking incredulous, 'Why would I do that?'

ANSWER

So, are you psychic? If you got 1/10 right, don't get too excited: the probability of this happening by chance alone isn't far off 50/50 (more precisely, it's 39 per cent). If you got 2/10 or 3/10, close, but no cigar: the probability of this happening by chance alone is 19 per cent and 6 per cent respectively. Things start to get interesting at around 4/10, as the probability of this happening by chance alone (1 per cent) is well under the conventional threshold of 5 per cent for 'statistical significance' (see Web Links for an explanation).

Beyond 5/10 (0.1 per cent) and 6/10 (0.01 per cent) the probabilities get so tiny that we need a better way to express them: for example, the number of copies of this book that I would have to sell before – on average – one reader gets this many right (assuming that everyone who buys the book takes this test once): 7/10 (1,143 copies sold) and 8/10 (27,435 copies) look achievable; 9/10 (1,111,111 copies) less so. In order to find a reader who got 10/10 right, I would, on average, have to sell exactly 100,000,000 – that's one hundred million – copies: enough to secure me a place in the Top Ten best-sellers of all time (right up there with *A Tale of Two Cities*, *The Lord of the Rings*, *The Hobbit* and *Harry Potter and the Philosopher's Stone*).

How do you compare to N'kisi? Well, out of 71 trials, N'kisi said 117 words (parrots obviously cannot be told to say only one word per trial!), of which 23 were hits. The probability of showing this kind of performance by chance alone is about 0.2 per cent, meaning that, in order to beat N'kisi, you would have had to have scored 5/10 or better. More importantly, since 0.2 per cent is well, well under the conventional threshold of 5 per cent, we are left with the somewhat uncomfortable conclusion that N'kisi has indeed shown evidence of paranormal behaviour. (One possible wrinkle is that a large proportion of N'kisi's successes involved his favourite word, *flower*, which he produced far more often than any other. However, even when these trials are excluded, his performance still comfortably passes the test for statistical significance.

So is N'kisi *really* psychic? Like most scientists, my instinct is to say, 'Of course not; he just got lucky'. Many would argue that the standard of proof required (i.e., setting the 'just got lucky' threshold at 5 per cent) is absurdly low, given how unlikely parrotnormal abilities are in the first place. But, as Sheldrake points out, most of us are happy enough to accept conventional scientific methods and statistical analysis techniques when they support our preconceptions, so it is unfair of us to cry foul when they don't. Let's dig a little deeper into this controversy with what – at first glance – appears to be . . .

Web Links

Statistical significance explained: https://benambridge.wordpress.com/companionsite/the-tea-test/

Calculate your own probabilities:
http://stattrek.com/online-calculator/binomial.aspx#faq

The best-selling books of all time.
http://en.wikipedia.org/wiki/List_of_best-selling_books

Sheldrake takes on the sceptics:
http://www.sheldrake.org/files/pdfs/Dogs_That_Know_Appx.pdf

A Shaggy Dog Story

Does your dog know when you're about to come home?

This was the question explored by Rupert Sheldrake in another investigation of animal extra-sensory perception. Of course, many dogs know roughly what time their owner usually comes home, or recognise the sound of the car engine, but are some dogs – even in the absence of these cues – able to just feel it in their bones when their owner's arrival is imminent?

Jaytee, a dog from Ramsbottom, near Manchester, developed a habit of going to the window every weekday at around 4.30 p.m., the time his owner, Pamela Smart, left work to begin her journey home. Nothing remarkable there, you might think. But then Pamela was made redundant, and, as a result, her comings and goings became more irregular. Nevertheless, Jaytee would still begin his windowside vigil at whatever time Pamela happened to be setting off for home (at least, according to Pamela's parents, with whom she lived).

Like N'kisi's owner, Aimée, Pamela decided to call in Rupert Sheldrake to conduct an investigation. Sheldrake also graciously invited the psychologist, ex-magician and professional sceptic Richard Wiseman to conduct an investigation of his own. Did Jaytee pass the test? You be the judge. The plot below (which summerises the data from Wiseman's first experiment) shows the length in seconds of Jaytee's visits to the window, along with the time and possible cause of each (because, of course, dogs look out of the window for plenty of non-psychic reasons too!). Does this pattern suggest that Jaytee sensed Pamela's return?

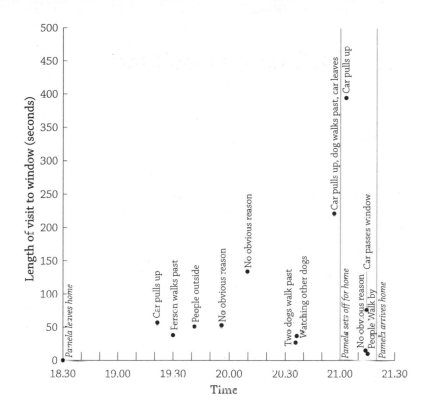

ANSWER

So, *Best in Show* or *Ain't Nothing But a Hound Dog*?

The problem is that it's all a matter of interpretation. Sceptics like Wiseman can point to the two window visits either side of 8 p.m. as evidence that Jaytee incorrectly 'guessed' this as Pamela's leaving time; and, although Jaytee made a particularly long window visit just after Pamela set off for home at 9 p.m., this was simply to look at a car that had just pulled up outside. Believers like Sheldrake can point to the fact that Jaytee put in far more window time in the two ten-minute slots during which Pamela was travelling home than in any of the other ten-minute slots throughout the evening; and it's hardly Jaytee's fault that by far his longest window visit – just four minutes into Pamela's homeward journey – happened to coincide with a car pulling up outside. After all, most times a car or person appeared, Jaytee watched for a minute or less, as compared with his six-and-a-half minute stint at 9.04 p.m. (The same argument can be made for all of Wiseman and Sheldrake's studies, as all of them – despite the researchers' disagreements over interpretation – pattern in a similar way.)

Do pets have psychic powers? You can make up your own mind, but I'm sceptical, and here's why: if parrots and dogs really are psychic, then why is their performance so imperfect? Why doesn't N'kisi guess the correct photograph every time (after all, if he's psychic enough to guess the photo, surely he's psychic enough to figure out the rules of Aimée's game). Similarly, why doesn't Jaytee just sit at the window for the whole of Pamela's return trip (surely what any dog who *really* knew that she was on her way home would do), rather than all this back and forth? So, if you scored 4/10 or better in the previous section, and are now going around saying that science has proved you're a psychic, ask yourself this: if you can really tell which picture your friend is thinking of, why did you get so many wrong?

Web Links

Sheldrake's take on his debate with Wiseman:
http://www.sheldrake.org/reactions/richard-wiseman-s-claim-to-have-debunked-the-psychic-pet-phenomenon

And vice versa (with an ingenious title, *How much is that doggy in the window?*):
http://www.richardwiseman.com/resources/Jaytee.pdf

Dogtanian and the Three Must-Get-Pairs

Whether or not dogs are psychic, they are certainly pretty smart. But who's smarter: dogs (technically the word refers just to males) or bitches (females)? Before we find out, let's have you compete against a member of the opposite sex on a task for which a large and reliable gender difference has been found (although you'll have to wait until afterwards to find out which sex does better). The task is the game Pairs, which you may well remember from your childhood (I had a *Postman Pat* version). You'll need to do a bit of preparation, but – hey – if you beat your partner, it will be worth it, right?

Cut out the twenty cards on the opposite page (or, if you prefer, print out the version available at www.areyousmarter-thanachimpanzee.com) and arrange them randomly, face down, on the blank grid.

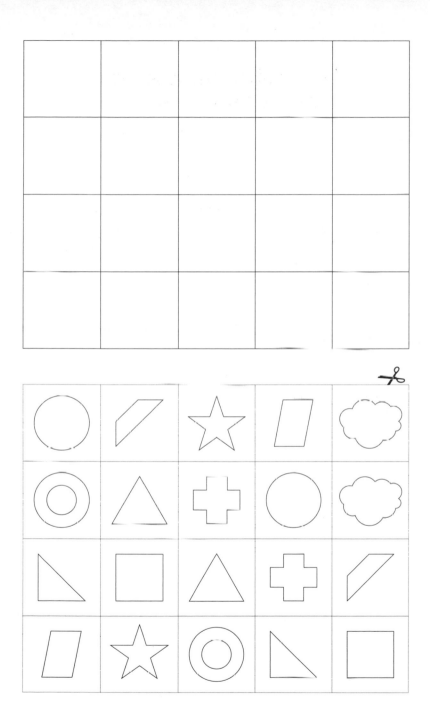

The game is played like this: pick two cards and turn them over. If you have a matching pair, remove both cards from the grid. If not, put both cards back, face down, in their original positions. Repeat until there are no cards left. That's it.

You and your opposite-sex opponent should play the game once each (with a different, and secret, random arrangement of cards) and note how many errors each of you make (an 'error' is simply turning over two cards that are not a pair). If you have a stopwatch handy, you could also time each other.

So, who won? When this game was played under test conditions, men made an average of 37 errors; women, just 25. What is more, men took an average of 3 minutes, 38 seconds; women 3 minutes, 27 seconds. Why did women beat men at this task? Well, the researchers actually predicted this result, on the basis that women have superior verbal abilities, and are therefore presumably better able to use a verbal strategy (e.g., saying to themselves 'the circle is in the top-left position'). However, this explanation doesn't seem to be right: the female advantage was not diminished when the familiar shapes were replaced with very unusual shapes chosen specifically to be almost impossible to name. This left the researchers rather stumped. They wondered aloud (well, in writing) about a link between higher oestrogen levels and better performance (as found in some previous studies), but – having failed to measure oestrogen levels – couldn't test this prediction directly.

This study piqued the interest of Charlotte Duranton, a French PhD student and dog lover (according to her web site, she owns a labrador, a Pyrenean shepherd and a French bulldog). Would female dogs outperform male dogs on a physical problem-solving task with a memory component (arguably) similar to the human task you just completed? The task was simple: a tasty treat (mince and chicken) was placed in a plastic box, which the dogs could open by lifting a plastic handle with their mouth or a front paw (although not all learned to do so). The memory component comes from the fact that each dog was given three goes at the test, with a 30-second wait between each go.

The findings were surprising, even to Duranton and her team. On the first run, roughly 65 per cent of male dogs succeeded, as compared with just over 20 per cent of female dogs. However, while the males showed little-to-no improvement on the second and third runs, the females' performance increased dramatically, with all succeeding on the third run. The male advantage on the first run is probably due to differences in boldness. While males and females seemed equally motivated to get the food, males were less worried about whether the box also contained any nasty surprises (see Left Behind? for a similar finding for humans). The female advantage for subsequent runs is probably due to the fact that – as in the study you completed – women seem to be better than men at remembering small-scale precise properties or features of objects; in this case, that the handle is the key to solving the task. Again, the reasons for this are a bit of a mystery, but the fact that a similar female advantage is observed in both humans and dogs suggests that the explanation is more likely to involve biology than sociology or linguistic ability.

In the meantime, here's one possible way to test the idea that a better memory for local details underlies women's superior performance on these types of test. If this is the case, then the female advantage in the pairs test that you completed should increase if the game is played multiple times with the pieces in the same starting positions. Why not try for yourself?

Turning Japanese

So dogs – perhaps especially female dogs – are pretty smart. But are they as clever as those legendary geniuses of the animal world, dolphins? Actually, probably not. We've seen (in A Walking Dogtionary) that dogs show a remarkable ability to learn words. Dolphins can go one stage further. They can understand sentences in which words are combined in new ways: i.e., sentences that they have *never heard before*. But can you? Let's find out by learning some Japanese.

Study the sentences below and their English translations:

Wani ga kirin-o ketta	The alligator kicked the giraffe
Zo-ga rakuda-o nameta	The elephant licked the camel
Kirin-ga kaeru-o butta	The giraffe hit the frog

Now, what does the following sentence mean?

Kaeru-ga Zo-o ketta

Do you need a clue? The key is to look through the three English sentences and find the only animal that appears twice.

Got it yet? 'Giraffe' is the only animal that appears in two English sentences, so if you look at the two corresponding Japanese sentences, you should be able to find the Japanese for 'giraffe': the key to solving the puzzle.

Still struggling? Well, once you know that the Japanese for 'giraffe' is *kirin*, you can figure out that the animal doing the action comes first and has *-ga* on the end (e.g., **Kirin-ga** *kaeru-o butta*, 'The **giraffe** hit the frog'), while the animal that has the action done to it comes second, and has *-o* on the end (e.g., *Wani-ga* **kirin-o** *ketta*, 'The alligator kicked **the giraffe**'). Armed with this knowledge, you can work out all the animal names, and that *Kaeru-ga Zo-o ketta* means 'The frog kicked the elephant'.

Two bottlenosed dolphins, Akekamai and Phoenix, were successfully taught a language that required a similar pattern of deductions. For example, two possible sentences in the dolphin language (presented using either hand signals or underwater clicks) were:

Pipe right-basket take	Take the right-hand basket to any water pipe
Right-pipe basket take	Take any basket to the right-hand water pipe

So, just as you learned the Japanese sentence pattern

[DO-ER(-ga)] [DONE-TO(-o)] [ACTION]

Ake and Phoenix learned the dolphin sentence pattern

[LOCATION] [THING] [ACTION]

A similar feat was achieved by Kanzi, a bonobo who was taught to produce and understand language using a touchscreen keyboard (see Web Link). Like Ake and Phoenix, Kanzi wasn't just learning words (which most dogs can do easily enough), but understood that the *order* of those words matters. For example, Kanzi could respond correctly to both *Put the hat on the ball* and *Put the ball on the hat* – sentences that use the same words, but in a different order.*

Why is this so important? Well, for many researchers, our use of rules of word order – and, with it, the ability to produce and understand completely new sentences – is the defining feature of human language, the characteristic that sets human language apart from the grunts and squeaks and squawks of the rest of the animal kingdom. So if dolphins and bonobos can do this too, this means that they too are capable of learning human languages, at least in a rudimentary form.

This begs an important question . . .

Web Link
See Kanzi in action on his language keyboard at:
https://www.youtube.com/watch?v=wRM7vTrIIis

---

* Kanzi's performance with these reversible sentences was by no means perfect; in fact, he was right just 57 per cent of the time. However, this beats chance performance (50 per cent) by a statistically significant margin (see Parrotnormal Activity for an explanation of statistical significance).

Why Can't We Talk to the Animals?

As a child, I suffered from a mild obsession with the film *Doctor Dolittle* (think Rex Harrison, not Eddie Murphy). At the heart of this obsession was a nagging question: *Why couldn't this be real?* After all, many pets seem to understand their own names and maybe a couple of simple commands, to say nothing of wonder-dogs such as Chaser (see A Walking Dogtionary). And on the speaking front, we have not only Alex and N'kisi (see Mathemalex and Parrotnormal Activity) but also Hugo, a German parrot who snitched on his owner's cheating husband, one Frank Ficker,* by repeating the name of his mistress (Uta). So what's stopping us? Why can't we go further and – like the eponymous doctor – hold conversations with our animal cousins?

Historically, a popular answer has been that non-human species lack the magical understanding that words can be put together in different orders to express different meanings. There's a saying in journalism that *Dog Bites Man* isn't news, but *Man Bites Dog* is. The joke (such as it is) makes sense only because we understand that the order of the words tells us who's doing the biting and who's getting bitten. But as we've just seen, some dolphins and bonobos can actually pass this test. Furthermore, there is some evidence that Alex the parrot (see Mathemalex) can even use this knowledge of word order to put together entirely new sentences. Similar claims have been made for chimpanzees' use of sign language, although this remains controversial (see, for example, the 2011 documentary film *Project Nim*, starring the chimpanzee Nim Chimpsky, named after the linguist Noam Chomsky).

---

* Yes, this is his real name. Obviously he wasn't quite as frank a Ficker as he could have been.

Thus, on one definition at least, some animals do seem to be capable of using 'language'. So why – in the main – don't they? If these animals are so smart, why aren't they explaining what it's like to be a chimpanzee, or at least politely asking to be let out of the cage?

Research conducted by Mike Tomasello, who has studied language-learning in both children and chimpanzees, suggests a new answer: what non-human species just don't seem to get is that language is fundamentally co-operative, almost altruistic, in nature. This is most obviously the case for warnings ('Look out, your boss is coming!'), where the language produced is entirely for the benefit of the listener, rather than the speaker. But even when the speaker's language is much more mundane (e.g., 'It's raining'), he presumably believes that the listener will find the information at least marginally interesting, and has incurred some small personal cost (i.e., the effort spent in producing the speech) to supply it. After all, nobody (well, almost nobody) provides a running commentary on unfolding events to an empty room. Sometimes, of course, it is the other way around: the listener is doing the speaker a favour by providing a sympathetic ear, perhaps at considerable cost (e.g., an hour of excruciating boredom).

Tomasello argues that this idea of doing something for the benefit of someone else, even if it involves no personal cost, is completely alien to chimpanzees. It would be anthropomorphic to call them 'selfish'. They don't consider the altruistic possibility and think, 'Sod that, I'm keeping all the bananas for myself'; they simply haven't evolved in such a way as to be capable of considering the altruistic option in the first place.

Evidence of this indifference comes from a number of laboratory studies. Many ape and monkey species respond to the presence of a predator by giving an alarm call, often interpreted as an altruistic warning to others. But a study with macaques showed that if a 'predator' (actually a lab technician in a surgical mask) approached a mother's baby in a different cage, the mother gave no alarm call unless she was also approached. In another experimental set-up, a chimpanzee was given the choice of two ropes to pull: one brought in five grapes for the puller and five for a chimpanzee in an adjacent cage; the other, five for the puller and none for his neighbour. Which option do you think the chimpanzees chose? The answer is that they pulled at random, neither deliberately feeding the neighbour, at no personal cost, nor depriving him. Again, not selfish; simply indifferent.* Human two-year-olds, when placed in a similar set-up, chose to feed a friendly adult, though only if she dropped hints ('I like crackers. I want a cracker'). Interestingly, eighteen-month-olds did not, suggesting that human concern for others may be learned rather than inborn. But either way, chimps don't show any.

Thinking about the bigger picture, virtually all of mankind's greatest achievements, such as science, government and the arts, are based fundamentally on co-operation. Or take money: bits of paper and metal and numbers on screens that have meaning only because we have collectively agreed to act as though they do. If humans really are qualitatively different from other animal species – and I have argued throughout the book that it is far from clear that this is the case – then an inclination to co-operation is perhaps the best candidate for that special something that makes us unique.

---

* But don't apes share food in the wild? A little bit, but usually only when they're trying to groom a potential coalition member or sexual partner. Sometimes they will tolerate another chimpanzee taking their food, but only when it is of very low quality. Similarly, although mothers feed their own children, they normally give them the peel or the shell and save the best bit – the banana or nut – for themselves. Can you imagine trying this with your kids?

Something To Crow About?

That said, Tomasello's view that co-operation is what sets humans apart from other animals is a new and relatively controversial one. An older – and perhaps still more mainstream – view is that man is unique in having *reason*: the ability to apply rules of logic, rather than simply following one's instincts. In fact, as we will see shortly, the jury is still out on the question of whether or not other animals are capable of logical reasoning. But are *you* . . .?

Let's imagine you are a naturalist studying a particular species of frog in the Amazonian rainforest. We know for sure that every member of the species is either green or orange (no frog has both colours). Your colleague has a theory about these frogs that he expresses in the form of the following rule (hint: the *precise wording* of the rule is *very* important):

If a frog is male, then it is green.

Now, you and this particular colleague have never really got on, so you have made it your mission in life to find one or more frogs that violate this rule and so disprove his theory (hey, naturalists can be just as petty as the rest of us).

1. You see a male frog. What must be true about the frog if it violates the rule?
   (a) The frog is green.
   (b) The frog is orange.
   (c) The colour of the frog doesn't matter: it can be green or orange and still not violate the rule.
2. You see an orange frog. What must be true about the frog if it violates the rule?
   (a) The frog is male.
   (b) The frog is female.
   (c) The sex of the frog doesn't matter: it can be male or female and still not violate the rule.

3. You see a green frog. What must be true about the frog if it violates the rule?
   (a) The frog is male.
   (b) The frog is female.
   (c) The sex of the frog doesn't matter: it can be male or female and still not violate the rule.
4. You see a female frog. What must be true about the frog if it violates the rule?
   (a) The frog is green.
   (b) The frog is orange.
   (c) The colour of the frog doesn't matter: it can be green or orange and still not violate the rule.

ANSWER

1: b, 2: a, 3: c, 4: c.

1. If you got this one wrong, then you really weren't paying attention. A frog that is both male and orange clearly violates the rule that 'If a frog is male, then it is green'.

2. This one is trickier, but only slightly. The same logic applies: a frog that is both male and orange clearly violates the rule that 'If a frog is male, then it is green'.

3. This one is much harder. If you said 'b', then you have fallen for the oldest trick in the (logician's) book. Remember, I told you to pay close attention to the precise wording of the rule. Your colleague said 'If a frog is male, then it is green'; he did NOT say 'If a frog is green, then it is male'. So a female green frog does NOT violate the rule that 'If a frog is male, then it is green'.

4. If you got this wrong originally, then the previous answer should have shown you the error of your ways: the colour of any female frog is irrelevant. She could be green, orange, purple or gold-plated, and still have no bearing on the truth or otherwise of the rule that 'If a frog is MALE, then it is green'.

Even if you didn't get all of these right (and the majority of people don't), the fact that at least some people get them right (and that most people can understand the answers once they're explained to them) demonstrates that we as a species are capable of logical reasoning.

Why is this important? Well, as we've already seen, logical reasoning is one of the best candidates for a uniquely human ability in its own right. But some researchers go further, and argue that our ability to perform logical reasoning depends crucially on the fact that we humans have *consciousness*: an even better candidate for that 'special something' that makes us human (though see Every Body Hurts?). There is some experimental evidence for this idea: getting people to perform a task that requires conscious processing (e.g., counting the number of times a particular word appears in a song) interferes with their ability to solve logic puzzles of the type you have just completed. This is consistent with the claim that our consciousness provides us with the awareness and reflection required to perform logical reasoning.

Viewed in this light, the question of whether or not any other species are capable of logical reasoning takes on a new level of importance. If they are, then we may have to credit them (reluctantly?) with consciousness (or, at least, something like it).

So are they? Animals cannot, of course, be given logic puzzles like the ones you have just done, so instead are given a simpler test of *reasoning by exclusion*. The animal is offered a choice of two cups, one of which has already been shown to be empty. If they are able to select the other cup, which contains a food reward, this is taken as evidence of reasoning (apes, capuchins, baboons, ravens and Clark's nutcrackers have passed the test). While the task sounds ridiculously trivial to us, the ability to infer that 'since this cup didn't contain the food, that other one must do so' is a form of logical reasoning.

That, at least, is the usual story. The problem is that it is actually possible to pass the test without doing this type of reasoning, simply by adopting a strategy of avoiding empty containers. It was not until 2015 that a group of enterprising kiwi researchers (the nationality not the bird) came up with an ingenious solution. In a study with New Caledonian crows, the cups were replaced with tubes.

In the crucial part of the study, both tubes are presented end-on, as shown opposite, with the food in the side of the bent tube. A crow that simply adopted a strategy of avoiding empty containers would either choose no tube at all or pick at random, as both appeared – on the face of it – to be empty. The only way to succeed is to perform bona fide *reasoning by exclusion*: 'I can *see* that the food isn't in the straight tube, so it must be in the side of the bent tube.' Most crows showed mixed performance, following this type of reasoning some of the time but also showing a general avoidance for (apparently) empty containers. However, two particularly cunning corvids displayed unambiguous evidence of the ability to perform reasoning by exclusion.

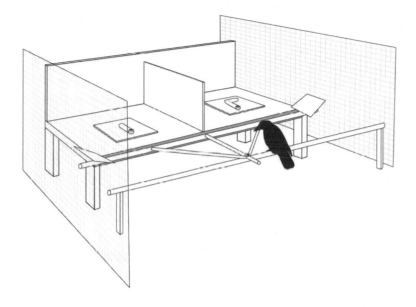

The fact that crows are capable of at least one kind of logical reasoning raises some profound questions. It is sometimes argued that this type of reasoning requires consciousness. So do we simply ditch that idea, or are we forced to concede that birds have consciousness too? And if birds have consciousness, then presumably all 'higher' creatures do too. Does this mean they have a *soul*? We will save an exploration of these difficult questions for the final section of the book, but in the meantime one thing is clear: the idea that only we humans can do logical reasoning is for the birds.

Students vs Squirrels, Sorta

But, say the human-exceptionalists, there is a still at least one particular *type* of logical reasoning that is unique to humans. Suppose, for example, that any box marked with this symbol

contains your favourite tasty treat. You are now shown six other boxes, each marked with a different symbol. Which three boxes do you think contain a treat, and why?

ANSWER

There's no right or wrong answer here; what we're interested in is the strategy that you used to make your choice. Did you pick the three symbols that just seemed most similar overall to the first symbol (the three in the bottom row, perhaps?), or did you follow an explicit logical rule based on shape (treats are indicated by stars, not circles), colour (by dark – not light – grey) or pattern (by dots, not the # pattern)?

This test, like the one in the previous section (Something To Crow About?), compares humans and non-humans on their reasoning abilities: specifically, how they sort things into categories. Although most of us have probably never thought about this much, there are basically two different ways to sort things (let's say, for example, clothes). The first is more intuitive: you just group together all the things that feel kind of similar in an overall way, in which case you might put shirts, vests and jumpers in one drawer and socks and pants in another. The second way to sort uses an explicit logical rule: for example, 'I'm going to put underwear in this drawer and outerwear in another' (in which case vests would go with socks and pants rather than shirts and sweaters) or even 'I'm going to put light colours in this drawer and dark colours in another'.

Humans are unique – so the theory goes – in our ability to create logical rules that are completely abstract (i.e., about shapes, colours and patterns) and arbitrary (there is nothing about a particular shape, colour and pattern combination that means 'the food is here'). The type of logical reasoning performed by the crows in the previous section – while impressive – is not nearly so abstract or arbitrary, as searching visually for food is something that they do every day. Arbitrary rules are so important to humans that we enforce them even when doing so goes against not only 'common sense' but even the wishes of almost all concerned. For example, it is not uncommon for criminal cases to be thrown out of court because a particular piece of paperwork was not submitted by an arbitrary deadline, even if the accused has already admitted his guilt.

The implication for the test that you just completed is that humans, perhaps especially university students (academia loves arcane, arbitrary rules), would be expected to rely on explicit logical rules, while – say – squirrels will rely more on the vaguer notion of overall similarity.

In fact, when given a version of this test, students and squirrels (and also pigeons) didn't differ: all three species were just as likely to use the supposedly more advanced strategy of picking one dimension (colour was the most popular) and sticking with it. If anything, humans were more likely than squirrls or pigeons to use the supposedly more primitive strategy of choosing on the basis of overall similarity.

What does this mean? Well, one possibility is that it is a mistake to equate the use of overall similarity with 'dumb' fuzzy thinking and the use of a rule as evidence of 'smart' logical reasoning. After all, it could be that using similarity rather than rules is a better strategy in everyday (human) life. This is the interpretation favoured by the researchers who conducted the study (which is unsurprising, as academics tend – on the whole – to be a rather sober and tentative bunch). However, they also concede that it is possible, on the basis of their results, that 'pigeons, squirrels and humans all have similar levels of access to analytic processes'. So, if you're a student, watch out: in your next tutorial you might just find yourself – if I may mix my animal metaphors – being outfoxed by a squirrel.

Box Clever

But, counter the human-exceptionalists again, there is yet a third level of reasoning beyond both the concrete and abstract reasoning that we met in the last two sections. I can't say any more at this stage without giving the game away, but you'll have to box clever.

In the pairs of boxes below, the one that contains the treat is marked with a *

Now, which box in the pair below contains the treat?

ANSWER

The answer is box (a). The correct strategy, which you should have been able to figure out from the first two examples, is to pick the lighter box in the pair. But I sneakily made this difficult by pitting this *relational* match against a *literal* match: the losing box (b) is actually the same colour as one of the previous winning boxes. Still, despite my trickery, you probably still got this right, as humans are very good at relational thinking.

But are other animals? We've already seen that some species can do concrete (**Something To Crow About?**) and even abstract reasoning (**Students vs Squirrels, Sorta**). But even this more abstract logical reasoning was based on *absolute* rather than *relative* characteristics: 'Is it the stars or the circles?' (not 'Is it the more-starlike shape?'); 'Is it the dot or the hashtags?' (not 'Is it the more-dotlike pattern?'). The reasoning required in the current section is even more advanced, because it requires you to pay attention to *relative* characteristics and to *ignore* absolute ones. Asking 'Is it the light grey?' will give you exactly the wrong answer. The right question is: 'Is it the *lighter* shade?'

The reason that this particular type of logical reasoning is potentially a last bastion of human uniqueness is that 'pick the lighter shade' seems like the kind of rule that needs to be explicitly verbalised, placing animals with no language (i.e., probably all of them*) at a distinct disadvantage. It would therefore be a real turn-up for the books if this ability were shown not by a primate or even a mammal, but by a humble reptile: say, a turtle.

Yep, that's right. This test was passed by two box turtles named Mario and Flippy, who – hold on, this gets even better – live at Walt Disney World and carried out the test in front of hundreds of adoring fans (OK, a handful of underwhelmed tourists who couldn't face the queue for Splash Mountain). Would you like to

---

* OK, some might *sort of* have language, or at least be capable of learning one, depending on our definition (see **Why Can't We Talk to the Animals?**), but none has the kind of language that lets you verbalise ideas like 'pick the lighter box'.

see a picture? Of course you would! Here's Flippy warming up before a test run (the food was hidden not in boxes, but behind paddles painted black, white or one of three shades of grey). Look, he's positively straining at the leash.

These tremendous *terrapene* were able to continue to pick the lighter shade, even when the black, white and grey paddles were replaced with ones in various shades of blue and green. This shows that they really had formed some kind of *general* concept of 'lighter' and 'darker', rather than a simple strategy tied to particular paddle pairs.

Now, if only the researchers had borrowed the animal participants from the previous section and trained them on ten times as many shades. *Then* I would have had a chapter title.* Still, the fact that Mario and Flippy could, without the aid of language, learn the relational concept of 'lighter' is not only flipping marvellous, but – I hope you will agree – turtley awesome!

---

* Have you got it yet? That's right – 50 *Shades of Grey Squirrel*.

Is There a Dog-tor in the House?

Mario and Flippy's home, Walt Disney World, is one of many locations where humans can swim with dolphins. Many people believe that animal-assisted therapy – swimming with dolphins, trekking with horses or simply playing with dogs – can be an effective treatment for all kinds of psychological disorders, from depression to post-traumatic stress disorder. But what does the evidence tell us?

Unfortunately, the answer is 'not much'. A recent review by animal psychologist Hal Herzog found that, although hundreds of studies have investigated animal-assisted therapy, the vast majority were not sufficiently well designed to allow us to draw any meaningful conclusions. A typical example is the field (pun intended) of equine (horse)-assisted therapy, which boasts no fewer than fourteen studies investigating its effectiveness – most with positive results. However, all but two of the studies had too few participants to allow us to be confident in the findings, and over half did not have a no-treatment control group. This is an important flaw, because many psychological problems resolve themselves over time anyway, meaning that we have no idea whether any improvement was due to the therapy. Perhaps most damningly, *not one* of these studies had a more sophisticated control group in which participants received *non*-animal-assisted therapy. So even if it turned out that people who received the therapy showed more improvement than those who didn't, we would have no idea whether or not it was down to the horses. For example, perhaps the handlers – who, let's face it, are the ones who actually speak to the participants – were behind any therapeutic effect.

The gold standard in scientific evidence is the randomised control trial (RCT), in which people are randomly assigned to receive either the treatment under investigation (here, some form of animal-assisted therapy) or some kind of control treatment (for example, otherwise similar therapy that does not involve animals).* But when a large group of Japanese researchers tried to conduct a meta-analysis – a special study that brings together the findings of all previous research on a particular topic – they found that it was simply impossible: there were just too few RCTs of animal-assisted therapy that met the basic standards for a well-designed experiment. In fact, regardless of quality, they were able to find just eleven RCTs – looking at dogs, cats, dolphins, birds, cows, rabbits, ferrets and guinea pigs – in total.

So, can a dog give you therapy? Who knows? But can you give therapy to a dog? Actually, yes. As any dog owner (or long-suffering neighbour) will attest, the most common psychological problem suffered by dogs is separation anxiety: many will bark, howl or chew (or worse) for hours on end, any time they are left alone. In the study in questions, in addition to an anti-depressant (Reconcile – essentially doggy Prozac) dogs were given behaviour-modification treatment that involved owners ignoring the dog for half an hour before leaving, giving a toy or treat on departure, and – on returning – rebuffing any excessive greeting and paying attention to the dog only when she calms down. The dogs that were put through this programme – but not a no-treatment control group – showed a significant reduction in all four measures of separation anxiety: inappropriate urination, inappropriate defecation, destructive behaviour and excessive vocalisation (i.e., barking or crying). In fact, the average (median) dog showed all four behaviours at the start of the study, but none after treatment.

Even better, the treated dogs did not just learn to mask their anxiety; they actually developed a more positive outlook. How do we know? Before and after the treatment, the dogs were given a

---

* The requirement that participants be assigned a treatment at random sounds like the height of pointless fastidiousness but is actually crucial: if people are free to choose their own treatment, it may well turn out that those who plump for animal-assisted therapy, rather than – say – a heavy dose of psychiatric medication, were less ill to begin with, rendering any comparison between the animal and no-animal groups meaningless.

cognitive bias test. This involves learning that, for example, a bowl on the dog's left always contains food, while a bowl on the dog's right is always empty. The speed with which a dog approaches a bowl in an ambiguous location (i.e., neither left nor right, but dead centre) is taken as a measure of her 'optimism'. Sure enough, the treatment group – but not the control group – became quicker at approaching this middle bowl, demonstrating that the training (or the doggy Prozac, or the combination of the two) resulted in a sunnier disposition (and, presumably, a lower neuroticism score on the canine personality test that we met in **A Stable Personality?**).

So, while the benefits of canine therapy for humans remain controversial, the benefits of human therapy for canines are well proven. But it's not just a one-way street. Did you know that dogs can detect cancer? I'd known about the claim for ages, but it seemed so unlikely that I never got around to checking it out. When I finally did, I was astonished to learn that dogs can be rapidly trained to detect, for example, lung cancer with 99 per cent accuracy just by sniffing people's breath.

What about the claim that dogs can detect human emergencies and seek help? Sadly, that one *is* a myth. In one study, dogs watched while their owners feigned a heart attack in an open field or were pinned to the floor by a fallen bookcase. Although a bystander was near by, not one dog approached her for help (although one dog – a toy poodle – jumped on to her lap and waited to be petted).

So, to sum up, while your dog might be able to sniff out cancer, don't rely on her in an emergency (even if she looks like the famous canine heroine opposite), or go to her for psychological therapy.

Let's Get Rat-Arsed

With animal-assisted therapy unproven, where are we poor humans to turn for help with life's trials and tribulations? One popular answer is 'to the bottle': we humans simply love to get off our faces on drink and drugs, sometimes even to the point that we become addicted. Why? Because we're animals. This behaviour is by no means restricted to humans: a recent *New Scientist* article (see Web Link) summarised some of the most heroic animal addicts, including wallabies on opium, shrews on booze, mice on speed, monkeys on marijuana and caterpillars on coke. But what if there were a simple test that could predict who is most likely to become addicted: one that could be taken by both humans and other animals?

Guess what? There is! Take the test below, then read on to compare yourself to the humble lab rat.

What would you rather receive in each of the following scenarios:

1. (a) £990 immediately OR
   (b) £1,000 in a week's time?

2. (a) £960 immediately OR
   (b) £1,000 in a week's time?

3. (a) £650 immediately OR
   (b) £1,000 in a week's time?

4. (a) £550 immediately OR
   (b) £1,000 in a week's time?

5. (a) £995 immediately OR
   (b) £1,000 in six hours' time?

6. (a) £960 immediately OR
   (b) £1,000 in six hours' time

7. (a) £920 immediately OR
   (b) £1,000 in six hours' time?

8. (a) £860 immediately OR
   (b) £1,000 in six hours' time?

ANSWER

This test is a measure of *delay-discounting*. People who show high rates of delay-discounting (i.e., of choosing a smaller reward in return for getting it sooner) are more likely to become addicted to alcohol, drugs, cigarettes, eating and gambling.* So, in general, the more 'b's you chose, the better. But, more specifically, what pattern marks you down as a potential junkie?

We can answer this question by looking at a study that gave these types of hypothetical scenarios to crack addicts and non-drug-users. On average, the non-users choose 'b' in every scenario except the first, while the addicts choose 'a' in all but 4 and 8. In effect, the addicts were happy to pay more than £300 to avoid waiting a week for their £1,000 and nearly £100 to avoid waiting just six hours.

How do you compare to a lab rat? In one study, rats were given the choice of 2 food pellets immediately or 12 pellets after a 15-second delay. (Rats, of course, cannot be given questionnaires regarding hypothetical scenarios, so were pre-trained that pressing one lever yielded the smaller immediate reward, and the other the larger delayed reward.) Nine of the 39 rats were unable to wait in 75 per cent or more of trials, and so were classified as impulsive. So, if you chose 'a' in any scenario except the first, congratulations: you're more impulsive than the average lab rat! Importantly, the rats identified as impulsive consumed more alcohol in a subsequent boozy buffet, and a later study repeated this finding with self-administration of intravenous cocaine. These findings are important as they suggest that, for humans, impulsivity (as measured by the type of delay-discounting questionnaire that you took above) is a cause, not a consequence, of drug and alcohol abuse.

After talking about these and related findings at a science festival, I was approached by a parent who was worried that her

---

* As we saw in my previous book, *Psy-Q*, they are also less likely to take preventative healthcare measures, such as wearing a seat belt, taking regular exercise, going to the dentist and even wearing sunscreen.

daughter displayed all the classic signs of an impulsive delay-discounter, and who asked if there is any kind of training that could help. Although this field is still in its infancy, a group of Arkansas researchers have recently begun to explore the possibility of using memory training. The idea is that the reduced ability to remember pleasurable experiences in the past and to envisage them in the future are two sides of the same coin, and that boosting the first should therefore boost the second. Early results are promising. The researchers indeed managed to reduce rates of delay-discounting among drug addicts, at least on the hypothetical questionnaire task. The £64,000 question, of course, is whether this training would ultimately be effective as a treatment for drug abuse.

And, if so, would it work on spiders? . . .

Web Link
http://www.newscientist.com/article/dn17373-animals-on-drugs-11-unlikely-highs.html#.VXBE-FxVhHw

Oh, What a Tangled Web We Weave

Another animal on which researchers have tested out various recreational drugs is the house spider. The reason that the spider makes a great, um, guinea pig is that drugs affect its web-spinning abilities . . . with hilarious consequences.

Let's see how you measure up. Below is a typical web. Your job is to try to copy this web (a) drug-free, (b) after a few cups of coffee and (c) after the drug of your choice (remembering to keep it safe, and keep it legal!) – perhaps alcohol. And, because simply drawing a web is a bit too easy compared with spinning one, let's have you use your wrong hand (i.e., your left hand if you're right-handed, and vice versa).

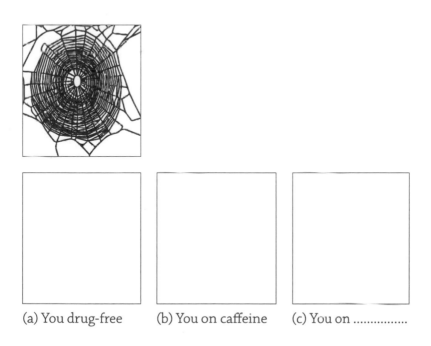

(a) You drug-free   (b) You on caffeine   (c) You on ....

ANSWER

Let's start by seeing how spiders get on. Below are the webs woven drug-free and on caffeine, marijuana, mescaline (peyote), sleeping pills (chloral hydrate) and speed (Benzedrine). Can you guess which is which (turn the page upside down for the answer)?

Unless you are of a very sensitive disposition, your on-caffeine web should look very similar to your drug-free web. Spiders are not so lucky. For them, caffeine is highly toxic: whatever drug you took when drawing your third web, I'd be surprised if you came up with anything nearly so crazy as the caffeinated spider. Indeed, if you tried to match the spider webs to the drugs on the basis of the severity of their effects in humans, you will have come up with almost exactly the opposite of the correct ranking.

Answers (from left to right): drug-free, marijuana, sleeping pills, speed, caffeine, mescaline.

The serious point here is not that drug studies on animals are a waste of time, as the results cannot be applied to humans. Rather, the lesson – as every animal experimenter knows – is that it is crucial to always choose the right animal model for the phenomenon or disease under investigation. Let's briefly explore a few examples (taken from the Royal Society's guide to the use of non-human animals in research).

As recently as the 1950s, polio was terrorising even the highly developed Western world. For example, a 1952 epidemic in the United States paralysed over 20,000 people and killed 3,000, most of them young children. An American researcher, Jonas Salk, famously took on the challenge of coming up with a vaccine. His approach involved deliberately infecting monkeys ('hosts') with the polio virus, then extracting the virus and killing it. This inactivated virus is then used as a vaccine. Following one of the largest clinical trials the world has ever seen – a double-blind trial with over 600,000 schoolchildren* – the vaccine was deemed both safe and effective, and Salk a national hero. Although, these days, the virus used to make the vaccine is grown in tissue culture rather than live animals, the vaccine would not have been developed in the first place without animal testing. Indeed, in some parts of the world (but not, since 2004, the UK) a live but weakened version of the virus is used, which requires each new batch to be tested on monkeys or – more recently – genetically engineered mice (in order to be sure that is sufficiently weakened).

Every year in the UK, 5,000 people develop kidney failure and require either dialysis or a transplant (without which, around a third of them would die). When, again in the 1950s, researchers began to investigate the possibility of transplants, they were not, of course, permitted to experiment on live humans. The only possibility was to use animals such as rabbits and dogs, which have

---

* 'Double-blind' means that neither the researcher nor the volunteer knows who is getting the treatment (here, the vaccine) and who an inert placebo, and is the gold standard of clinical trial research.

similar respiratory (breathing) and cardiovascular (heart and blood-flow) systems to humans. Early efforts were frequently hampered by the immune system's rejection of the transplanted organ. So the researchers went back to the rabbits and dogs, and came up with a successful immune-suppressant drug. Even patients who do not receive a transplant owe their lives to animal testing: these patients must undergo dialysis, which requires an anti-clotting agent discovered in research on dogs.

We could go on: treatments for cancer, epilepsy, cystic fibrosis . . . almost every major medical advance has made extensive use of animal testing. And this is true not only for the 'physical' diseases discussed above but also for 'psychological' conditions. For example, animal models are used for the development and testing of anti-depressant drugs. If a rat is placed in a water-filled tube from which escape is impossible, it will swim, struggle or try to escape for a while, and then give up – moving into the state of 'despair' that is analogous to depression in humans. A simple and effective way to test whether or not a particular drug might have anti-depressant properties in humans is to investigate whether it staves off this 'depression' in rats (i.e., whether it increases the time before the rat gives up and stops struggling). Indeed, as we saw in Is There a Dog-tor in the House?, anti-depressants developed on rats for humans often work for other species too.

Some opponents of animal testing argue that differences between other animals and humans mean that the results cannot be generalised from one to the other. The examples above, and many more that I could have chosen, demonstrate that this is not the case, provided that the species is carefully chosen to be similar to humans in the relevant respects (e.g., to have similar respiratory and cardiovascular systems when investigating kidney transplants). Even apparent counter-examples, when examined more carefully, in fact demonstrate the crucial importance of animal testing. For example, it is often noted that the Thalidomide disaster, which caused babies to be born with malformed limbs, was not anticipated by animal testing. But this is only because researchers did not investigate this possibility by giving the drug to pregnant animals. When, belatedly, these animal studies were conducted, the same effects were observed.

Neither are we, yet, in anything like a position to be able to replace all *in vivo* testing (in living animals) with *in vitro* testing (testing 'in glass': i.e., test tubes, flasks or Petri dishes containing tissue culture), or with computer simulations (and even for the narrowly circumscribed areas where we *have* reached this point, it has generally taken many years of animal testing to get there). There are still a great many cases in which we can properly evaluate the effects of a drug or procedure only in whole living bodies.

You will have your own view on the ethics of animal testing, and it is not for me to seek to impose mine on anybody else. Some people believe that the life of a human is worth no more than the life of an animal, and that we should not kill even a single mouse to save a human life (although the calculations are rarely this straightforward – what about killing several thousand monkeys to save tens of thousands of human children, as in the case of the polio vaccine?). While perhaps unusual, there is nothing illogical or unscientific about this viewpoint: it is a matter of personal ethics rather than logic and science. But let's not kid ourselves: given our current state of medical knowledge, if we were completely to abandon animal testing tomorrow, thousands of humans would die of potentially treatable conditions.

Web Links
Feed drugs to spiders and watch them spin at:
http://www.badspiderbites.com/spiders-on-drugs/

The position statements of the UK Royal Society and the US National Academy of Sciences can be found at:

https://royalsociety.org/~/media/Royal_Society_Content/policy/publications/2004/9726.pdf
http://books.nap.edu/openbook.php?record_id=10733

Insex

Putting such morbid thoughts aside and returning to our industrious spider, the reason that he spins a web in the first place (hopefully before, rather than after, his morning coffee) is to catch insects. Fortunately for the hungry spider, there are plenty of them about. Insects are prodigious and enthusiastic breeders. In fact, some of the things they get up to are downright kinky. Some of their more outlandish sexual practices are listed below. Your job is to (a) pick out the ones that are real from the ones where I'm pulling your leg (of which every insect has six) and – because the idea of this book is to pit yourself against our animal friends – (b) tick off any that (assuming that you could!) you might be prepared to attempt yourself after a few glasses of wine and a candlelit dinner.

	True	False	Sign me up!
The male's penis explodes as he ejaculates.			
The male has two penises, but mainly uses only one, keeping the other as a spare in case the first one breaks off.			
The male uses his spiny penis to create small tears in the vagina.			
Foreplay consists of the two insects firing mucus-covered 'love darts' at each other.			
The male uses his dagger-like penis to inject his sperm into the female through her skin. The eggs then hatch inside the mother, and the babies eat her alive from the inside out.			
These tantric sex masters can keep going for up to 79 days (move over, Sting).			
The male mates with both females and other males.			
This hermaphrodite insect not only fertilises its own eggs with its own sperm but also creates sperm-producing organs inside its children, which in turn fertilise them.			
Two hermaphrodite insects fence with their penises. The loser is inseminated through its skin and becomes the 'mother'.			
The female gobbles up the male's sperm ... using an extra stomach attached to her vagina.			

ANSWER

All are true, except the ones about the love darts and the penis-fencing; and these are false only because the animals that do engage in them (snails and flatworms) are not insects. While many of these sexual practices might strike us humans as amusing (or just downright painful), there is a serious side: each one improves the insect's chances of success at the mating game and/or harms the chances of its competitors.

For example, the honeybee's exploding penis isn't an evolutionary 'mistake'. Rather, the explosion leaves behind a plug that acts as a barrier against rival sperm. Similarly, male stick insects keep at it for days and weeks on end, not because they are tender and considerate lovers, but because – to put it bluntly – when one penis is in there, no more can get in. And if, like the earwig, you have a penis to spare, you don't even need to hang around: you can simply leave one in there as a contraceptive and get on with your day.

Quite a few insects mate by simply injecting the female with sperm, presumably because this is quicker and easier than the method we mammals use, although *strepsiptera* (parasites that live inside wasps and other larger insects) are unique in eating their mothers alive. Even insects that mate via the vagina can, like male bean beetles, benefit from having a penis that punctures and tears the skin; the wounds allow the ejaculate, which contains a fertility-boosting chemical, to enter the female's body cavity. This is not, however, the purpose of the female butterfly's extra vaginal stomach. On the contrary, this ' bursa copulatrix' allows her to devour the nutrients in the male's sperm package while neutralising the fertility chemicals. Why would she want to do that? Well, the longer she goes without getting pregnant, the more nutritious sperm packages she receives.

Hermaphroditism is not unusual in the insect world, but having children to which you are both father and grandfather certainly is. It is the innocent-sounding cottony cushion bug that achieves this feat, by leaving in its offspring organs that create its own sperm. Same-sex sexual behaviour is even less unusual among insects, with over a hundred species dabbling. However, a recent review suggested that at least the majority of encounters may be cases of mistaken identity. Males are more often mounted by other males when they are carrying female pheromones (picked up in hetero-sexual encounters) and often resist. In other cases, the male being mounted encourages the other by mimicking the behaviour – or even pheromones – of the female, apparently to reduce aggression from a dominant male or to distract a rival from an opportunity to mate with a female.

You might think that snap-off or exploding penises, weeks-long intercourse, vaginal stomachs and all the rest of it are a million miles away from human sexual experience. But actually, many of our sexual characteristics presumably serve the same functions. What, for example, is the point – evolutionarily speaking – of sexual jealousy? The point, for males, is that it motivates them to stave off rivals, just as surely as a vaginal plug or left-behind penis.

So, odd as they may seem, most of the more colourful aspects of insect sexuality do seem to serve a serious purpose. Quails, on the other hand, are just downright kinky . . .

Let's Spend Some Quaility Time Together

It's true! Quails – you know, the birds with the eggs that taste exactly like chickens' eggs, but are smaller and therefore somehow 'fancy' – can develop sexual fetishes that would put Christian Grey to shame. But what about you? Take the test below to find out how your fantasies compare to those of a group of middle-aged Berliners.

How arousing do you find each of the following *as a fantasy* (even if you would never dream of acting it out)?

	Not at all arousing	Slightly arousing	Moderately arousing	Quite arousing	Very arousing
Transvestitism (cross-dressing)					
Fetishism (use of non-living objects: e.g., shoes)					
Voyeurism (secretly watching others while they are naked or undressing)					
Exhibitionism (exposing yourself to an unsuspecting stranger)					
Frotteurism (rubbing up against an unsuspecting stranger)					
Masochism (being tied up, humiliated or beaten)					
Sadism (tying up, humiliating or beating others)					

ANSWER

So how do you compare against our sample of middle-aged German men. The percentage of men who ticked 'slightly arousing' or more for each of these categories is as follows:

Transvestitism 5 per cent; Fetishism 30 per cent; Voyeurism 35 per cent; Exhibitionism 4 per cent; Froteurism 13 per cent; Masochism 15 per cent; Sadism 22 per cent.

So, you now know whether or not you're kinkier than the average German male; but what about the average Japanese quail? In the study in question, male quails were given the opportunity to mate with a female, having been briefly shown a piece of terrycloth (that stuff used to make dressing gowns, bathrobes and so on) immediately beforehand. After thirty runs through this procedure, roughly half of these males had developed a 'fetish' for the piece of terrycloth: that is, they attempted to mate with it.

Finally, in the test phase all males were given the chance to mate with their familiar (real-life quail) partner but also with the terrycloth in the cage too. The males who had developed a fetish were slower to mount the female than those who had not, apparently because they were too busy eyeing up (or more!) the terrycloth. But when they did finally get around to it, they successfully fertilised a greater proportion of the eggs. What was the cause of this rather unexpected result? Well, as the researchers rather primly put it: 'It may be that the increased time and attention that the fetishistic male quail directed toward the conditioned stimulus object [the terrycloth] helped to prime their reproductive physiology.' In other words, the fetishistic males got so turned on by the terrycloth that they ejaculated more – or harder – than their more strait-laced counterparts.

What is the point of all this? Well, psychologists have long thought that unusual human fetishes might result from the accidental pairing of – for example – a shoe and a sexual experience (i.e., via Pavlov's-dog-style classical conditioning). But this hypothesis is almost impossible to test in humans because: (a) we can't study them until they're adults (by which time their proclivities are probably already pretty well developed); and (b) we can't offer them sexual experiences in a laboratory (other than very mild ones such as viewing pornography). Although human sexuality is almost certainly rather more complex than quails', the findings of this and similar studies do suggest that the Pavlovian approach is at least along the right lines. After all, if humans can get aroused by dressing themselves as stuffed animals (autoplushophilia), by stone and gravel (lithophilia) and by being cold (psychrophilia), is it really so surprising that a quail will try to mate with a dressing gown?

And if you think that's a sick idea . . .

A Sick Idea

What are the next two numbers in this sequence?

2   3   5   7   11

ANSWER

If you are struggling, you should know that a humble insect has solved this problem.

Does it help if I tell you that this insect is the cicada?

The reason it might is that cicadas are famous (relatively speaking) for having a life-cycle that is a prime number.*

That's right: 13 and 17.

Cicadas spend almost their entire lives (13 or 17 years, depending on the species) underground. They then emerge, all at once, for a final hurrah: a month or so in the sun, during which they sing (loud!), mate, lay their eggs and die. When the eggs hatch, the nymphs immediately go underground and the cycle begins again.

The advantage of having a long life-cycle is that predators (mostly birds) will be overwhelmed. Since the cicadas emerge all at once, there are still plenty left to reproduce even after every bird has eaten as many as it possibly can. The advantage of having a prime-number life-cycle is that the cicada is able to avoid predators synchronising their life-cycle with its own. For example, if cicadas had a life-cycle of fifteen years, a bird species that had a life-cycle of 5 years could synchronise its cycle so that, every three generations, its newly hatched chicks – and hence its peak population – would experience an unusually large bounty of cicadas. But by maintaining a cycle of 17 years, the cicada ensures that these cycles will coincide only every 85 years (the 17-year cycle of the cicada multiplied by the 5-year cycle of the bird).

At least that's the usual story. But actually, perhaps the best-supported hypothesis is that prime-number cycles prevent cicadas from synchronising not with birds but with themselves. If cicadas from broods with different life-cycles were to mate, then their offspring would have a life-cycle with an unusual length. This would be disastrous, as these offspring would emerge when there

---

* If your high-school maths is a little hazy, you may need a reminder that a prime number is any number greater than 1 that cannot be divided exactly (i.e., without leaving a remainder) by any other number (except, of course, by itself and 1).

are no other cicadas to mate with (and plenty of ravenous birds to eat them). By maintaining large, prime-numbered life-cycles (i.e., 13 and 17 years), cicadas can ensure that broods with different life-cycles will meet only every 221 years (13 × 17).

Whether or not the evolutionary purpose of prime-number life-cycles is to avoid predators, there's a much more common trick for doing so that is used by creatures great and small. Can you guess (hint: humans do it too)?

You've Got To Hide Your Bug Away

That's right, camouflage. Given the widespread use of this tactic in the animal kingdom, it's amazing how long it took us humans to hit on it. For example, the British Army 'Redcoats' did not abandon their distinctive brightly coloured uniforms until 1902 (perhaps because of the psychological advantages for the wearer, see Ape, Man United), and only then for head-to-toe khaki. The camouflage that we associate with the army did not become widespread until the 1960s. Evolution worked out the advantage of camouflage much earlier. For example, imagine you're a hungry bird. Can you spot a tasty snack in the picture below?

OK, so you probably managed it eventually, but it was pretty hard, right? (And, if anything, it's even harder in the colour version; see the companion web site.) Certainly a hungry bird flying past would be lucky to spot it. But look what happens if we flip the caterpillar over:

It sticks out like the proverbial sore thumb, doesn't it? Why is the same caterpillar much more difficult to spot in the first picture than the second? The answer is a phenomenon known as *counter-shading*. Many animals – including whales, dolphins, tigers and tuna fish – are darker on top and lighter underneath. As well as providing protection against the sun, this is often a very effective camouflage strategy. If you take an object that is a uniform colour all over and hold it in the sun (or under an artificial light), you will notice that the top appears much lighter than the bottom. This means that an animal whose skin or fur was, in reality, the same shade all over would be easy to spot against a uniform background (say, grass, the desert, the sky or the sea), as its top half would be lit up like a Christmas tree. Countershading counteracts this effect. If, in terms of its skin or fur pigmentation, the animal is darker on top and lighter on the bottom, then sunlight from above will lighten the darker top half to the same shade as the genuinely lighter bottom half. As a result, the animal will appear to be a uniform shade all over, and can camouflage itself against a background of the same colour.

The caterpillar that you just saw (or rather, didn't see) is the exception that proves the rule. Unlike whales, dolphins, tigers and tuna fish, it is *lighter* on the top than the bottom. This is because caterpillars of this particular species (*Actias luna*) prefer to hang upside down. So, when in their preferred position, the top half (the belly) is darker than the bottom half (the back), and counter-shading makes for very effective camouflage (as in the first picture). But if you flip the caterpillar over (as in the second picture), the pattern of sunlight and shade makes his lighter back look even lighter, and his darker belly even darker, making him extremely easy to spot.

So if Eric Carle's famous caterpillar wants to satisfy his own hunger, rather than that of the birds, I've got a tip for him: get on your back!

Should I Stay or Should I Go?

When your prey uses cunning mathematical and camouflage techniques to try to evade you, it's a good idea to have a trick of your own up your sleeve. Fortunately, many animals do. But do you? The following test applies, in principle, to any hunting or foraging scenario, but – since most of us humans no longer have to track down our own dinner – we'll use something more genteel: apple-picking.

On the next page is an orchard. Before looking at it in detail, use a blank piece of paper to cover up all but the two leftmost trees. Now toss a coin to pick which of these trees will be your starting tree.

Ready? Your job is to harvest as many apples as possible from this orchard in five hours. These five hours are divided up into fifteen twenty-minute slots, each represented by a tick box. You should tick off each slot as you use it. In each slot you can either: (a) harvest the number of apples shown in the highest as yet unticked box on that tree (by ticking the relevant box); or (b) move on to a new tree (by ticking the box on the 'travel' arrow and tossing a coin to pick your next tree). For example, assuming that your toss came out heads for your start tree, you could collect:

1. 64 apples, then move on
2. 64 + 32 apples, then move on
3. 64 + 32 + 16 apples, then move on
4. 64 + 32 + 16 + 8 apples, then move on
5. 64 + 32 + 16 + 8 + 4 apples, then move on
6. 64 + 32 + 16 + 8 + 4 + 2 apples, then move on
7. 64 + 32 + 16 + 8 + 4 + 2 + 1 apples, then move on

Because moving on to a new tree costs you one of your fifteen blocks in travelling time, you are faced with a constant dilemma: *Should I stay*, and face ever diminishing returns on this tree, *or should I go*, and forfeit twenty minutes of apple-picking time?

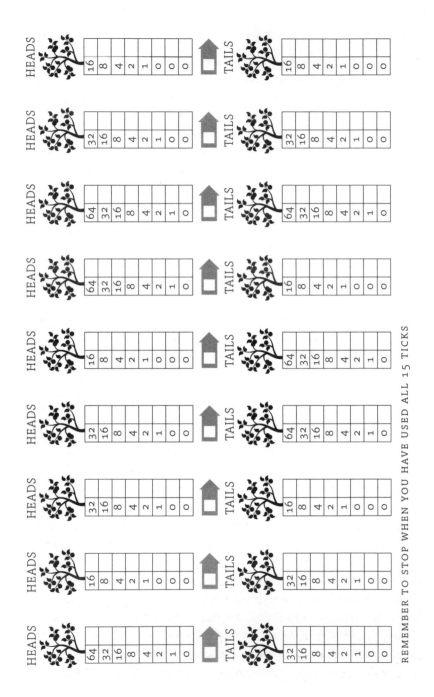

REMEMBER TO STOP WHEN YOU HAVE USED ALL 15 TICKS

ANSWER

How many apples did you get? The best possible score is 448, although this can be achieved only if every single coin flip is in your favour. A more realistic target is 300, which – unless you are a very unlucky coin tosser – you should be able to get by refusing to hang around for fewer than 16 apples per slot.

The optimal strategy in these *patch-foraging* tasks is to move on to the next patch whenever the current option is less than the *opportunity cost,* calculated as the long-run average per slot. For example, suppose that after spending ten time slots (some on harvesting, some on travelling), you have collected 96 apples. This means that a time slot is worth 9.6 apples. So, if only 8 apples are on offer in a given slot, collecting them is, quite literally, not worth your time; you would be better off investing that time in moving on to a more fruitful patch.

Although few people have heard of this *marginal value theorem*, most of us are pretty good at applying it intuitively, at least in these kinds of study. We're not perfect, however. A large-scale computerised version of this task found that, on average, people harvested roughly 10 per cent fewer apples than they could have done, owing to over-harvesting: sticking around when they should have moved on. Interestingly, the study turned up a few extreme over-harvesters: people who *hardly ever* moved on to a new tree, sticking with the current one all the way down to zero. This has interesting implications for real-life *optimal stopping* problems. At what point, for example, should you cut your loses on your current job, investment or marriage and move on? The answer, if the findings from the apple-harvesting task translate into the real world, is 'sooner than most of us do'!

Others animals have, of course, not heard of the *marginal value theorem* at all. But this doesn't stop them applying it more effectively than many humans. Animals whose foraging behaviour closely matches the predictions of the theorem include rhesus monkeys (who can be trained to do similar computerised tasks to humans), guinea pigs, hairy armadillos and various bird species.

So when the time comes to renew your car insurance, and you can't face shopping around for the best deal, you could always ask a blue tit.

Great Tits and Seedy Locations

From blue tits to great tits. These birds have a particularly impressive ability: they are brilliant at remembering the location of hidden food. But are you?

The grid below shows 100 hiding locations, five of which house a sunflower seed. Study the grid, then set an alarm on your phone for an hour's time (or, if you really want to test yourself, twenty-four hours) and place a bookmark on the following page, which shows a blank version of the grid. When you come back, try to fill in the five 'seedy locations' on the blank grid.

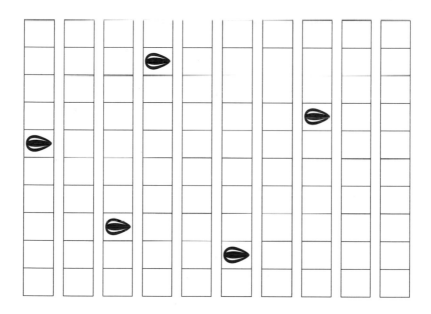

Great Tits and Seedy Locations: Test

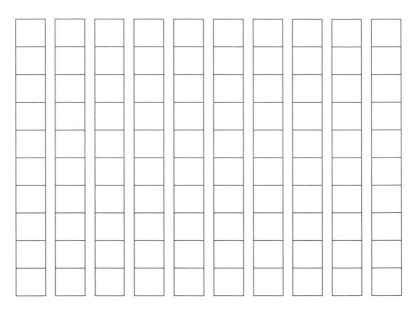

Great tits scored an average of 30 per cent with a one-hour delay and 20 per cent with a twenty-four-hour delay (measured as the percentage of the first ten searches that were successful). The experimental set-up consisted of ten artificial trees, each with ten hiding holes drilled in the trunk (each hole was covered with a piece of cloth so that the birds couldn't see the food).

Did you beat them? If so, don't be too pleased with yourself, as you have two important advantages. First, great tits don't know that the 'aim of the game' is to find the seeds with as few searches as possible, so may just store the rough location of each seed and search in all of the holes that are there or thereabouts. Second, you can use language to encode the location of each seed (e.g., 'third row in, three up from the bottom', 'eighth row in, seven up from the bottom') and even, if you're particularly smart, store this information using mnemonics (e.g., 'Uncle Arthur lives at number 33 [3,3], so I'm going to picture him eating a sunflower seed, sitting on the 87 [8, 7] bus).' Great tits, on the other hand, must rely on visuo-spatial memory alone, which makes their performance particularly impressive.

But what's special about great tits? After all, lots of birds – including crows, jays, nutcrackers and chickadees – are 'hoarders', who store food in times of plenty, and rarely forget where they left it. The answer is that great tits are not hoarders but thieves. Their special ability lies in remembering where *other birds* left their stash, and then pilfering from it. (In the memory test described above, great tits didn't hide the seeds themselves, but watched members of another bird species doing so.) So, the reason that great tits' memory abilities are particularly impressive is that they don't share the special brain adaptations enjoyed by hoarders (including a larger hippocampus), which allow them to hide and retrieve food instinctively. Great tits must develop this rather devious strategy for themselves.

One might expect, therefore, that some great tits will be better at this task than others. And so it turns out. A follow-up study by the same authors found that females outperformed males by a large margin (approximately 40 per cent vs 15 per cent 'hits' in the first ten searches). Indeed, unexpectedly, female great tits were just as good at finding the seeds as members of the hoarding species who had hidden them in the first place. This female advantage is surprising given that, for most species, it is the males who outperform the females on tasks involving visuo-spatial memory.*

So what's going on? The answer seems to be that female great tits are better thieves because they need to be. Males are dominant over females and elbow them out of the way (if birds have elbows) whenever a natural food source is available. If they are to avoid starvation, then, females must make use of an alternative 'artificial' source: other birds' stashes. Even in hoarding species –

---

* This includes humans. You can read more about the (small) male advantage in visuo-spatial abilities in my previous book, *Psy-Q*.

such as willow tits – highly dominant members hoard far less food than subordinates, safe in the knowledge that, if needs must, they can always bully their way to a good meal. And that is why, in birds, this dominance hierarchy is often called the 'pecking order'.

The Pecking Order

The term 'pecking order' (or, in the original German, *Hackordnung*) originated from research on chickens. Most animal species that live in groups have a pecking order, or dominance hierarchy, which governs who gets the first access to resources such as food and mating partners. Having a pecking order, in which everyone knows his or her place, means that individuals don't have to fight it out every time an opportunity for feeding or fornicating arises. There is still the odd fight every now and then, when a young upstart decides to try to move his way up the pecking order, but these are usually relatively mild or even just a ritualised display (e.g., showing off one's size, teeth or roar). For example, chickens establish the pecking order by clawing, jumping on or (of course) pecking one another, with the subordinate usually submitting (by running away or crouching) before serious injury occurs. They can also learn by watching confrontations between others: a hen will rarely challenge a stranger who has just knocked a formerly dominant individual off her perch.

As we saw in A Stable Personality?, when it comes to dominance, humans are the exception that proves the rule. Although we can all picture 'alpha-male' types who rule the roost in many social situations, human inventions such as money, social class and academic reputation mean that, in the right scenario, even the wimpiest geek can find himself in a position of dominance. Consequently, it is not really possible to measure dominance among humans at the individual level.* One thing we can measure, however, is *social*

---

* People have certainly tried. For example, both the California Psychological Inventory and the Jackson Personality Research Form contain dominance scales. However, perhaps because of the difficulty in tying down the concept, dominance has never really made it into the mainstream personality inventories (see A Stable Personality?).

dominance orientation; the extent to which people would like their own ethnic, national or religious group to dominate others (i.e., the extent to which they value a 'pecking order' between groups, rather than individuals). You can measure your own social dominance orientation by filling in the questionnaire below.

Which of the following objects or statements do you have a positive or negative feeling towards? Beside each object or statement, tick the box that represents the degree of your positive or negative feeling.

	very negative	negative	slightly negative	neither positive nor negative	slightly positive	positive	very positive
1. Some groups of people are simply inferior to other groups.	1	2	3	4	5	6	7
2. In getting what you want, it is sometimes necessary to use force against other groups.	1	2	3	4	5	6	7
3. It's OK if some groups have more of a chance in life than others.	1	2	3	4	5	6	7
4. To get ahead in life, it is sometimes necessary to step on other groups.	1	2	3	4	5	6	7
5. If certain groups stayed in their place, we would have fewer problems.	1	2	3	4	5	6	7
6. It's probably a good thing that certain groups are at the top and other groups are at the bottom.	1	2	3	4	5	6	7
7. Inferior groups should stay in their place.	1	2	3	4	5	6	7
8. Sometimes other groups must be kept in their place.	1	2	3	4	5	6	7

	very negative	negative	slightly negative	neither positive nor negative	slightly positive	positive	very positive
9. It would be good if groups could be equal.	7	6	5	4	3	2	1
10. Group equality should be our ideal.	7	6	5	4	3	2	1
11. All groups should be given an equal chance in life.	7	6	5	4	3	2	1
12. We should do what we can to equalise conditions for different groups.	7	6	5	4	3	2	1
13. Increased social equality.	7	6	5	4	3	2	1
14. We would have fewer problems if we treated people more equally.	7	6	5	4	3	2	1
15. We should strive to make incomes as equal as possible.	7	6	5	4	3	2	1
16. No one group should dominate in society.	7	6	5	4	3	2	1

To find your social dominance orientation score, add up the numbers in all of the boxes that you ticked, and divide by 16. If your score is over 4, then you have some tendency towards valuing social dominance. Most people have much lower scores. For example, one study of students at a top American University (albeit probably a rather liberal, egalitarian sample) had a mean of just 1.74. Another study compared scores across twenty different countries. The ranking, from the highest need for dominance to the least was: Serbia, UK (the shame!), Indonesia, Taiwan, USA, Switzerland, Italy, New Zealand, The Netherlands, Turkey, Northern Ireland, Ireland, China, Lebanon, South Africa, Spain, Belgium, Greece, Bosnia-Herzegovina, Poland. How embarrassing:

we Brits even 'beat' the USA – a country that we love to mock for being, as we see it, excessively patriotic.

Moving swiftly on, perhaps even more interesting is the original study, which looked for correlations between individual participants' scores on the social dominance questionnaire and measures of their personality, ideology and views on social policy. Higher scores on the social dominance questionnaire were associated with lower levels of concern for others, communality, tolerance, altruism, *noblesse oblige* (the idea that we should help people who are poor or lower-ranking), support for social programmes, racial equality policy (including affirmative action), women's rights (including equal pay), gay and lesbian rights, environmental programmes and mixed-race couples. Higher scores on the social dominance questionnaire were associated with higher levels of anti-Black racism, nationalism, sexism, equal opportunities, patriotism, cultural elitism, political-economic conservatism, support for military programmes, chauvinism, voting Republican, a tough stance on law and order, decreased immigration and the death penalty. Phew!

So if, as a human, you were feeling superior about our lack of dominance hierarchies, these findings should give you pause for thought. At least at the group level, many members of our species are in favour of a pecking order.

A Weighty Problem

Sticking with chickens, here's a classic teaser.

A farmer is driving a van full of chickens to market. The combined weight of the van, driver and chickens is 4,400 pounds. But, oh no, the famer has to drive across a weak bridge, which bears the sign: TOTAL WEIGHT MUST NOT EXCEED 4,000 POUNDS. The farmer has a brainwave. He has 100 chickens in the back, each of which weighs 5 pounds. So, flooring the accelerator, he speeds towards the bridge. At the very last moment, he hammers on the back of the cab. With a squawk, the startled chickens fly up into the air. With the total weight of the van now just 3,900 pounds, the farmer makes it across the bridge just before the chickens land. Phew!

Would this work?

ANSWER

No. And, before you ask, no, it doesn't make any difference if the van is sealed (except that the chickens would probably be dead, rendering the question moot).

Birds fly by pushing air downwards, effectively increasing the weight of the van. In fact, a study with parrots found that, on the downstroke, birds push air down with a force almost double their body weight.

Is there even a glimmer of hope for the farmer? Kind of: the study found that birds generate almost no downward force on the upstroke. So, in principle, if the farmer could time his dash across the bridge so that all or most of the chickens were simultaneously on the upstroke, his ruse would work. In practice, the chickens would all flap at different times, meaning that the average weight would remain roughly constant.

If this is all a bit too theoretical to convince you, you may like to watch the episode of the US TV show *Mythbusters* in which they tried this experiment with pigeons (see Web Link).

Web Link
Mythbusters: Bird in a box (watch from 16:40) https://www.youtube.com/watch?v=jmoq5CVohqg

For Eagle-Eyed Readers

Whatever was going on in the back of that farmer's van, the chickens probably had a pretty good view of it. A little-known fact about chickens is that they have excellent colour vision. While we humans have only three different types of 'cone' cell in the brain (red, green and blue),* chickens have five (red, green, blue, violet and black-and-white), arranged in a high-density mosaic pattern.

Although not widely known, the fact that chickens have excellent vision shouldn't really surprise us. After all, chickens are birds (if slightly odd ones), and pretty much all birds have better eyesight than humans (hence the cliché of the 'eagle-eyed reader', or having 'eyes like a hawk'). There are many reasons for birds' superiority over mammals in the vision stakes. For one thing, birds have more photo-receptors (like having a camera with more megapixels). For another, their eyes have very soft lenses, which can change shape quickly and easily, allowing birds such as kingfishers to see equally clearly through air and water. Another advantage is that the back part of the eye, on to which the image falls, is much flatter in birds' eyes than our own, which are almost round. To appreciate the difference, imagine if the flat screen in your local cinema were replaced with the inside of a giant football. A small area in the middle would just about be in focus, but the rest – the periphery – would be blurred beyond recognition.

Is our peripheral vision *really* that bad? It certainly doesn't feel like it as we go about our everyday business. Except at the very periphery, somewhere around your ears, it *seems* as if pretty much everything is in focus. Actually, this is an illusion, caused by the fact that we rapidly move our eyes so that whatever we are looking

---

* Actually, a few lucky people have four: http://www.popsci.com/science/article/science/woman-sees-100-times-more-colors-average-person?TCxHoGoOZkWM-MoGX.03

at falls right in the middle, with the brain doing some clever processing work to ensure we don't notice. In fact, our peripheral vision is terrible, as you can see for yourself with a quick and simple exercise (taken from Daniel Dennett's *Intuition Pumps and Other Tools for Thinking*, an excellent book that I can't recommend highly enough).

Don't look now, but on page 182 is a life-size picture of a playing card (or, if you have one handy, you can use a real card). Hold your non-card hand in front of you at the distance you would normally hold a card, in order to read it comfortably. Now, keep looking at your hand and slowly move the card until it covers your hand. Do NOT move your eyes! You need to be very careful to avoid the overwhelming temptation to sneak a look at the card as it comes in. If you have the self-discipline to pull this off, you will discover that you won't be able to tell the value of the card, or even the suit, until it is almost directly in front of you.

Fowl Play

What do you see here?

A grey diamond, in front of a black square, right?
Now, which line is longer, A or B?

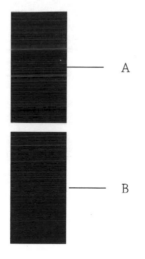

ANSWER

Actually, the two lines are the same length, but most people think that A is longer. Because we can't see the left-hand edge of Line A, we assume that it continues behind the rectangle. Of course, at a conscious level you know that this isn't true. You're looking at a two-dimensional page, for goodness' sake. I couldn't make the line go behind the rectangle any more than I could make it leap up off the page. But this illusion can't be reasoned away: the tendency to perceive objects as behind or in front of one another is completely automatic. Consider, for example, the black 'square' on the previous page. What if I told you that, actually, it's not a square but this:

It doesn't make any difference, does it? When you look back to the previous page, you still see a grey diamond in front of a black square; your brain fills in the 'hidden' part automatically. In general, this *completion* phenomenon is a good thing. If it didn't happen, you'd be very confused whenever objects passed in front of one another in the real world, which they tend to do quite a lot. The price you pay is that when one object (e.g., Line A) really does happen to end just where another (e.g., the grey rectangle) begins, you risk being misled.

Chickens show a different, and rather puzzling, pattern. In the study in question, two bantams (Bizen and Chris) learned to classify lines as either 'long' or 'short', by pecking at one of two response keys for a food reward. Then, in the test phase, grey rectangles were added, either touching the line (A) or with a gap of a few millimetres (B). Reversing the pattern shown by humans, the chickens were more likely to classify a line of a particular length as 'short' if it touched the rectangle (A) than if it did not (B).

Although nobody has been able to come up with a satisfactory explanation of why these chickens seemed to show the illusion in the 'wrong' direction, the fact that they don't show the human pattern of thinking Line A is longer than Line B suggests that their brains – unlike ours – don't automatically complete hidden objects. So are we smarter than chickens because we fill in the missing details that are (usually) there? Or are they smarter than us, because they're not fooled by a 2D image on a screen or printed page?

Whatever the answer, as we saw in the last section (For Eagle-Eyed Readers) there can be no doubt that the visual systems of chickens – and birds in general – are very impressive. But what's perhaps even more impressive is that some animals can 'see' without using their eyes at all. And so, with a bit of practice, can you . . .

For Eagle-Eyed Readers: Test

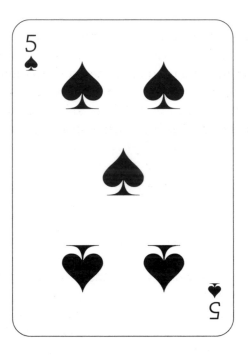

Are You a Bat, Man?

D. C. Comics' Batman had no special powers beyond his astounding intellect and limitless wealth, but real bats have a genuine superpower: they can 'see' in the dark, using what's called *echolocation*. This involves making a series of chirps and listening for the echo that bounces back from the objects near by (see Web Links for a video of a bat using echolocation to catch its prey). The system is incredibly flexible. When cruising for insects,* the bat chirps at a slower rate, which gives it a range of 15–20 metres. But when honing in on its prey, the bat ups its chirping rate by a factor of 20, narrowing its range to less than 1 metre, but ensuring that it receives an update on the precise location of its prey at least 10 times every second.

Not all bat species echolocate. With a few exceptions – including the Egyptian fruit bat – the larger *megabats* do not use this trick; it is mostly the preserve of the smaller *microbats*, including the infamous blood-sucking vampire bats.

Oh, and humans. A number of blind people have trained themselves to use clicks to 'see' their way around the environment in pretty much the same way as a bat would. Indeed, a recent brain-scanning study found that these real-life Batmen and women didn't have abnormally good hearing; rather, the part of the brain that usually processes input from the eyes had learned to process input from the ears instead.

So let's try. Many experts say that the best way to begin to echolocate is to take a drive with the windows down. Listen to the way objects such as buildings and parked cars 'whoosh' by as the sound from your car bounces back off them, but then take note of

---

* Bats use echolocation not only for catching insects but also for navigating their way around busy caves in the dark; see, for example, http://www.bristol.ac.uk/news/2015/october/obstacle-avoidance-by-bats.html

whether you hear such noises when you drive through an empty field. Notice, too, how when you pass rounded objects, such as trees or telegraph poles, the noise fades in and out as the sound waves bounce back in all directions. For a flat object, such as a phone box, the noise is shorter and of a constant volume, as the waves bounce back only when you are at exactly 90 degrees to it. This, in its most basic form, is echolocation.

Now let's try clicking. There are many different ways to click, including snapping your fingers or even using a manual clicker; the important thing is to settle on a sound that you can make exactly the same every time. Perhaps the most popular method of clicking is to place your tongue against the cavity in the roof of your mouth and pull it quickly backwards. The noise is made when the middle of the tongue (not the tip) comes away from the palate, releasing the suction. Be sure not to hit any other part of your mouth with your tongue, and keep your mouth wide open to ensure a high-pitched click. This is important because sounds with a low pitch are not very directional (i.e., it's hard to tell where they're coming from); exactly what you don't want when trying to echolocate.

Here comes the difficult bit: seeing with echolocation. Stand in an empty field (or, failing that, a large, empty car park), in order to avoid competing echoes bouncing back off buildings, and hold a football about ten inches in front of your face. Start clicking, at a rate of two to three clicks per second, and listen to the sound as it bounces back off the ball. Now try the same thing with a large bowl. You should notice that the sound coming back off the inside of the bowl is louder and contains more different frequencies than the sound coming back off the football. This is because the concave shape of the bowl means that it is very efficient at 'scooping up' the sound and reflecting it straight back at you (in just the same way as the concave shape of a satellite dish – or a radio telescope – focuses all the incoming signals on to the receiver). In contrast, the convex surface of the football means that the sound bounces off and disperses in all directions.

When you think you're ready, have a friend hold up either the ball or the bowl without telling you which, and see if you can tell the difference. Do this ten times. If you get at least 9/10 right, then congratulations, you're an echolocator!*

Of course, learning to echolocate effectively takes years of dedicated practice. But for blind people, the potential rewards are huge: those who put in the hours can use the technique to navigate their way around the world completely unaided. For example, one particularly remarkable video (see Web Links) shows blind echolocator Daniel Kish happily riding his bike around the neighbourhood.

As Kish points out, many people are deeply sceptical about human echolocation. But the science backs it up. Furthermore, one study showed that blind echolocators experience some of the same 'visual' illusions as sighted people. Take the size–weight illusion. For all of us, small objects feel heavier than larger objects of identical weight (try it for yourself). Because we expect small objects to be light, when we encounter a small but heavy object, the sensation of 'Woah, this is heavy!' causes us to overestimate its weight. This study found that blind echolocators show the same effect when they 'see' large and small objects of identical weight using the clicking technique.

Finally, back to bats. Did you know that, despite the well known phrase, bats aren't actually blind? Their eyes work perfectly well; it's just that they tend to sleep through the day, and come out to hunt only when it gets dark (which no doubt contributes further to their sinister public image). So if someone calls you 'blind as a bat', take it as compliment; it actually means 'not blind at all, but able to see both by day and by night'.

Web Links
The tutorial in this section is based on Tim Johnson's free online introductory course:
http://www.humanecholocation.com/introduction-to-echolocation/

---

* Why 9/10? Well, you could expect to get around 5/10 correct even if you were just guessing, and 9/10 is the minimum score that is sufficiently greater than this at-chance level to pass the statistician's test of 'statistical significance' (see Parrotnormal Activity).

Here you can see Daniel Kish using echolocation as he rides his bike; more information can be found on his web site:
https://www.youtube.com/watch?v=A8lztr1tu4o
http://www.worldaccessfortheblind.org/

Here's a bat using echolocation to catch a moth: https://www.youtube.com/watch?v=po8YooRAX3g

An Elephant Never Forgets

Speaking of which, have you heard the parable of the blind men and the elephant? A version of this story is part of the mythology of all the major Indian religions. A number of blind men are asked to touch, and then describe, an elephant. Each explores a different part (e.g., leg, tail, tusk, trunk etc.) and so comes up with a completely different description (e.g., it's like a tree, rope, spear, snake etc.). What's the moral? Well take your pick: it could be anything from the need for communication and respect for different viewpoints to the possibility of being simultaneously right and wrong, and the wave-particle duality (light, for example, behaves like particles in some ways, but like a wave in others). Thus the story is its own metaphor: everyone has a different interpretation, but none is unambiguously correct on its own.

If this is making your head spin, you may like to ponder instead an alternative elephant aphorism: *An elephant never forgets*. This is not only a lot simpler, but pretty much true. Well, OK, they probably forget *some* stuff; but elephants do have incredible memories. For example, in one study, researchers tried to trick elephants by spraying the urine of their family members in various locations. When they came across a patch that shouldn't have been there ('Sue's been over there all morning – I haven't seen her around here'), the elephants showed signs of surprise. The researchers concluded that elephants are able to remember not only thirty different individuals, but also where they are – a feat that puts humans to shame (imagine going on a trip with thirty friends or family and knowing at all times where everyone is).

Another large mammal with a large memory is the panda. One recent study found that Long Hui, a male panda at Vienna zoo, could remember a simulated panda face when tested a year later. The findings of the study suggest that pandas may recognise one another by – cuteness alert – their distinctive eye masks. But can you? Earlier in the book, you saw a picture of Gao Gao, a panda at the San Diego zoo. Can you pick him out of the pair below (turn the page upside down for the answer)?

Meanwhile, back to elephants: did you know that some always twist their trunks clockwise, others, always anti-clockwise? It's true. But why...?

Gao Gao is on the right.

ANSWER

A Shrewd Judgement

Here's a simple (and painless) experiment to try on your cat. Simply hold up a treat or toy so that she begs for it, or tries to swat it with her paw. Now repeat the experiment several times, in different locations and with different treats or toys. All will be revealed in a second, but for now, simply make a mental note of the position of your cat's body as she reaches (or take photos).

If you don't have a cat, we can run the experiment on you: go to the kitchen and get yourself a biscuit. Pick up the biscuit and eat it. Done.

Now it's time to make the titular shrewd judgement: do animals have a hand (or paw) preference? That is, while you probably always use the same hand when you're reaching for a tasty treat, does your cat? And if you repeated the experiment with a northern tree shrew (with mealworms as the treat – yum!), would he or she always use the same hand?

ANSWER

As recently as 1980, handedness was widely thought to be a uniquely human trait. But, as for so many of the human traits discussed in this book, the slow drip of scientific findings has gradually eroded away our claims to uniqueness. We now know that a preference for one limb over the other is shown by species as diverse as octopuses, toads, turtles, birds, pigeons, chickens, parrots, bats, mouse lemurs, dogs, elephants (where, charmingly, it manifests as a preference for twisting the trunk one way or the other) and many species of ape and monkey. Oh yes, and cats. And northern tree shrews.

The similarities don't end there. For humans, the degree of preference for one hand over the other isn't fixed but depends on the environment. For example, when we can't see the object we're reaching for (e.g., because the light is poor), we almost never use our non-preferred hand. When plenty of visual information is available, we're quite happy to use our non-preferred hand, if the situation merits it (e.g., I just picked up my coffee cup with my left hand, because I was using my right hand to move the mouse). Several monkey species (including brazza and squirrel monkeys) also show this pattern, although tree shrews – chosen for study because this species is unusually closely related to primates – do not (this may sound surprising, but actually tree shrews are not really shrews at all). This suggests, then, that flexible, visually-modifiable handedness seems to have evolved at some point after primates sprouted their own branch on the evolutionary tree of life.

So is there *anything* unique about human handedness? One possible candidate is population-level handedness. Somewhere in the region of 90 per cent of humans describe themselves as right-handed. But if you take a pandemonium of parrots, a destruction of cats or a meadow of shrews* (see School of Collective Nouns), each group will be split roughly 50/50 between right- and

---

* All right, I made this one up, in honour of Shrewsbury Town, who play at Greenhouse Meadow (having recently moved from the flood-prone Gay Meadow).

left-handers (or, to use an Americanism that seems particularly appropriate here, the north- and southpaws). As recently as the turn of the millennium, you could have made a pretty decent case that population-level handedness is indeed unique to humans. But in the past decade and a half it has become increasingly clear that – yet again – our claims to uniqueness are almost certainly mistaken. Chimpanzees, in particular, show pretty convincing evidence of population-level handedness (roughly two-thirds are right-handed, although it depends on the test used).* On the other hand (sorry!), bonobos – tied with chimpanzees for the distinction of being humans' closest living relatives – do not.

This isn't just an idle curiosity; the difference between chimpanzees and bonobos hints at an explanation for why it is that humans are predominantly right-handed. According to one theory, human right-handedness evolved hand-in-hand (sorry again!) with tool use. The brain systems that control tool use are found mainly in the left hemisphere. Because the left hemisphere controls the right hand (and vice versa), an increase in tool use could plausibly be linked with an increase in right-handedness. A rival theory has it that human right-handedness evolved together with language. Again the brain systems responsible – for both spoken language and gesture – are found mainly in the left hemisphere. Thus an increase in the use of language could also plausibly be linked with an increase in right-handedness.

Which of these two rival theories is right? Well, the fact that chimpanzees are predominantly right-handed while bonobos are not supports the tool-use theory. Why? Well, bonobos are generally thought to surpass chimpanzees in terms of their linguistic abilities (Kanzi, who we met in Why Can't We Talk to the Animals?, is a bonobo), while chimpanzees have the edge in terms of tool use. So, the fact that we humans are more similar to chimpanzees than to bonobos in terms of our right-handedness suggests that it is linked to tool use, rather than language. In fact, chimpan-

---

* Another (perhaps unlikely) candidate is the kangaroo, most of whom seem to be left-handed: http://www.bbc.co.uk/news/science-environment-33169547

zees themselves show evidence of right-handedness when using tools in the wild, including when leaf-sponging (using a leaf as a sponge), ant-dipping (using a shoot as a spoon to pick up ants) and termite-fishing (using a thin twig as a rod to catch termites).

So let's give a right-handed salute to our ingenious primate ancestors who – by simply trying to reach a tasty snack – influenced the way that we humans do the same thing some seven million years later.

Left Behind?

So, human handedness can tell us a lot at a population level; but what can it tell us at the individual level? Could there be links, for example, between handedness and personality . . .?

In A Stable Personality? we met perhaps the most popular system for classifying the personalities of humans and other animals: the OCEAN model, in which individuals vary on Openness to experience, Conscientiousness, Extraversion, Agreeableness and Neuroticism. However, an alternative system – Jeffrey Alan Gray's bio-psychological theory – also has many enthusiastic supporters. For its advocates, a particularly attractive feature of Gray's approach is the way that it seeks to explain personality biologically, in terms of the relative dominance of two competing brain systems. Although sceptics would no doubt see this as biological reductionism of the very worst kind, the evidence for Gray's approach is – as we will see shortly – rather compelling. So, what are these two systems? Before we find out, let's measure them in – you guessed it – you.

Each item of this questionnaire is a statement that a person may either agree with or disagree with. For each item, indicate how much you agree or disagree with what the item says. Please respond to all the items; do not leave any blank. Choose only one response to each statement. Please be as accurate and honest as you can be. Respond to each item as if it were the only item. That is, don't worry about being 'consistent' in your responses.

	Very false for me	Somewhat false for me	Somewhat true for me	Very true for me
1. A person's family is the most important thing in life.	1	2	3	4
2. Even if something bad is about to happen to me, I rarely experience fear or nervousness.	4	3	2	1
3. I go out of my way to get things I want.	1	2	3	4
4. When I'm doing well at something, I love to keep at it.	1	2	3	4
5. I'm always willing to try something new if I think it will be fun.	1	2	3	4
6. How I dress is important to me.	1	2	3	4
7. When I get something I want, I feel excited and energised.	1	2	3	4
8. Criticism or scolding hurts me quite a bit.	1	2	3	4
9. When I want something I usually go all-out to get it.	1	2	3	4
10. I will often do things for no other reason than that they might be fun.	1	2	3	4
11. It's hard for me to find the time to do things such as get a haircut.	1	2	3	4
12. If I see a chance to get something I want, I move on it right away.	1	2	3	4
13. I feel pretty worried or upset when I think or know somebody is angry at me.	1	2	3	4
14. When I see an opportunity for something I like I get excited right away.	1	2	3	4
15. I often act on the spur of the moment.	1	2	3	4
16. If I think something unpleasant is going to happen, I usually get pretty worked up.	1	2	3	4
17. I often wonder why people act the way they do.	1	2	3	4

	Very false for me	Somewhat false for me	Somewhat true for me	Very true for me
18. When good things happen to me, it affects me strongly.	1	2	3	4
19. I feel worried when I think I have done poorly at something important.	1	2	3	4
20. I crave excitement and new sensations.	1	2	3	4
21. When I go after something, I use a 'no holds barred' approach.	1	2	3	4
22. I have very few fears compared with my friends.	4	3	2	1
23. It would excite me to win a contest.	1	2	3	4
24. I worry about making mistakes.	1	2	3	4

First off, you can disregard your responses to questions 1, 6, 11 and 17. These are 'fillers': irrelevant questions designed to throw you off the scent if you were trying to figure out how the questionnaire worked. Fillers play an important role in many psychology experiments, but are generally all too rare in personality tests.

Add up your scores for questions 2, 8, 13, 16, 19, 22 and 24 to measure the strength of your **Behavioural Inhibition System (BIS)**. The average score for the BIS is around 18 for men and 21 for women. This brain system is involved in avoiding things that are boring, painful, new and scary, and otherwise unpleasant. People with a high score on the BIS are particularly sensitive to such unpleasantness and go out of their way to avoid it. They also show particularly high levels of anxiety when escape from such a situation is difficult or impossible. For example, one study found that participants who scored high on the BIS showed higher levels of anxiety than lower BIS-scorers when asked to undergo a painful but harmless procedure (keeping their hand in an ice bath for one minute) as a punishment for failure on a learning task.

The yin to the BIS's yang is the **Behavioural Approach System (BAS)**. This brain system is involved in seeking out things that are satisfying, fun or otherwise rewarding. The BAS can be broken down into three separate components. Add up your scores for questions . . .

- 3, 9, 12 and 21 to measure your **drive** (the average score is around 12, with no difference for men and women)
- 5, 10, 15 and 20 to measure the extent to which you are **fun-seeking** (the average score is around 12, with no difference for men and women)
- 4, 7, 14, 18 and 23 to measure the extent to which you are **motivated by reward** (the average score is around 17 for men and 18 for women).

People with a high score on the BAS are driven and motivated, and show higher levels of positive emotions such as hope and happiness, particularly when they are working towards a goal. For example, one study found that participants with high scores on the BAS scales rated themselves as happier than lower BAS scorers during and after (but not before) a learning task in which they could win extra points for good performance.

A particularly interesting aspect of the BIS/BAS approach to personality (or, as I like to call it, the *BISh-BASh-bosh* approach: 24 questions and you're done) is that the idea of a battle between two rival systems isn't just a metaphor. Rather, the inhibition/avoidance and approach/reward systems (the BIS and BAS respectively) are real-life brain systems whose activation we can measure using scanning techniques such as Magnetic Resonance Imaging (MRI). When we do so, it seems that the BIS and BAS are housed mainly in the right and left hemisphere respectively. For example, one study found that participants who showed a greater degree of left than right hemisphere activation in a certain part of the brain (pre-frontal scalp regions) scored higher on the BAS (the good one) and lower on the BIS.

This leads to an intriguing possibility: because being right-handed reflects a dominant left hemisphere and vice versa, we might expect handedness to be linked to personality, with right-handers scoring higher on BAS traits (e.g., drive, reward, fun-seeking) and left-handers scoring higher on BIS traits (e.g., anxiety and avoidance). This possibility is quite difficult to test with humans, partly because we are pretty good at hiding any anxiety we may be feeling, and partly because university ethics committees take a dim view of placing participants in genuinely threatening, anxiety-provoking situations.

One solution to this problem is to test marmosets: small monkeys that can quite easily be placed in unfamiliar – or downright terrifying – scenarios. When this is done, it turns out that – yes – left-handed marmosets are indeed more anxious than right-handers. They are less willing to approach, sniff or taste new foods and freeze for longer when they hear a hawk (all right, a recording of one, played by the mean experimenters).

Another way to sidestep the practical difficulties associated with terrifying human volunteers is simply to look for associations between handedness and BIS/BAS scores. A complicating factor is that, unlike marmosets, humans don't divide neatly into right- and left-handers: many people are inconsistent, preferring their left hand for some tasks and their right for others. Nevertheless, on the whole, the findings of studies that have investigated this question suggest that – yes – left-handers score higher on the BIS than right-handers, while the pattern is reversed for at least one element of the BAS: fun-seeking. But let's not get carried away. As is almost always the case for these types of study, the difference between right- and left-handers – though statistically significant – was very small. So the next time you meet a left-hander, don't immediately assume that she is an anxious, timid introvert. Unless, that is, she is a marmoset.

Cuckoo Clocks His Rivals

One thing that many humans – left- and right-handers alike – are perennially anxious about is losing their partner to a rival. For heterosexual men, perhaps an even more serious concern is being *cuckolded*: unwittingly tricked into raising a child who is the product of another man's loins. This topic has been the focus of a good deal of research, not just because psychologists are interested in other people's sex lives,* but also because the similarity between the behaviour of humans and other species once again casts light on sexual selection (see Spot the Difference and How the Giraffe Got His Neck) and, therefore, on how each of us came to be here today. To find out how, the male partner in your relationship should fill in the questionnaire below (I'll leave it up to you whether or not you share your answers).†

1. On a scale of 1 (not at all) to 10 (extremely)
   (a) How physically attractive do you currently find your partner?
   (b) How sexually attractive do you currently find your partner?
   (c) How physically attractive do other men currently find your partner?
   (d) How sexually attractive do other men currently find your partner?

Take the average of these scores (i.e., add them up and divide by 4) to find your partner's overall **attractiveness** score.

---

* That said, avoiding cuckoldry is probably one of the main reasons why humans have evolved to love gossiping about who is sleeping with whom.
† The original study was conducted on monogamous heterosexual couples only. So if you're in a different type of couple, by all means fill it in, but be aware that the results won't necessarily apply.

2. How many (a) male friends and (b) male work colleagues does your partner have?

Add these two numbers together to find your overall number of **sexual rivals**.

3. How many times did you and your partner have sex (with each other!) in the last week?

ANSWER

Before delving into your sex life, let's find out what happened when psychologists at Oakland University gave this questionnaire to almost 400 young men (with an average age of twenty-four) who had been with their partner for, on average, around three years.

First, the researchers divided the men into those with more attractive partners (an average attractiveness score of around 9/10) and less attractive partners (an average of around 7/10; unsurprisingly, almost nobody gives their partner 5/10 or lower). This allowed the researchers to discover an interesting pattern: on average, the more male friends and colleagues a woman has, the more often she has sex with her partner; but this is true only for the women in the 'more attractive' group. For the women in the 'less attractive' group, there was no relationship between the number of male friends and colleagues and how often she has sex with her partner (or, as the researchers romantically put it, 'in-pair copulation frequency').

Let's put some numbers on this to see how you compare. For women in the 'more attractive' group (around 9/10), those with fewer male friends and colleagues (up to 37) had sex with their partners an average of just under three times per week. Those with more male friends and colleagues (37 or more) had sex with their partners an average of just over four times per week. Women in the 'less attractive' group had sex with their partners an average of around twice per week, irrespective of the number of male friends and colleagues. Relationship length matters too: on average, each year of the relationship 'costs' you around 1.5 copulations per week.*

What is going on here? The answer favoured by the researchers (hey, don't shoot the messenger) is sperm competition: men

---

* Obviously these types of averages have to be interpreted with caution as, no matter how long you've been together, even the most hopeless lover can never have fewer than zero copulations per week.

with very attractive partners are 'worried' that their partners are having sex with other men. The more men she encounters in her daily life, the greater the risk that – there's no way to put this gently – she is carrying around their sperm, and thus the greater the need for her partner to, ahem, give his own sperm the best possible chance in that competition. The reason 'worried' appears in 'scare quotes' is that, in the modern world, most men in committed relationships can be pretty confident that their partners are not going to have sex with male friends and colleagues the minute their backs are turned. They are 'worried' only in the sense that evolution has hard-wired into men the desire to have even more sex if their partner is particularly attractive and there are a lot of potential rivals on the scene.

This may sound like a load of sexist claptrap, as it seems to imply that men have sole control over how often each couple has sex. As the researchers acknowledge, another possibility is that it is the attractive women who are initiating all this sex, having had their desire piqued by attention from male friends and colleagues (although arguably this is even more sexist, as it implies that women find attention from male friends and colleagues a turn-on, rather than annoying).

In fact, the original explanation – that this 'extra' sex is a result of men's evolutionary insecurities regarding sperm competition – does not imply that men have the only – or even the major – say in how often each couple has sex. All it implies is that men have *at least some* say in the matter, which, of course they do (for example, by deciding whether to accept or rebuff their partner's advances). Whether or not you are willing to accept this sperm-competition hypothesis for humans, it is almost certainly true for other species, such as wild fowl: a study published in the none-more-prestigious journal *Nature* found that as the number of competitors (male 'audience members') increased (from zero, to one, to three), so dominant males increased the number of sperm that they ejaculated in each copulation. A similar study found that human males produce more mobile sperm when masturbating to pornography featuring two men and one woman (i.e., a sperm-competition scenario) than to otherwise similar pornography featuring no men and three women.

So, guys, if you and your partner are trying for a baby, now might be the perfect time to bring up that threesome you've always had at the back of your mind. The only potential downside is that the third party needs to be a virile-looking man. You're cool with that, right?

Dad Calls It Quits

Ultimately, at least in evolutionary terms, the reason than men engage in all this rather icky-sounding sperm-competition business is to produce one or more offspring. But having done so, who makes a good dad, and who clears off at the first sight of a dirty nappy? To find out, let's turn the tables and have women complete the first part of the questionnaire from the previous chapter.

On a scale of 1 (not at all) to 10 (extremely):

(a) How physically attractive do you currently find your partner?
(b) How sexually attractive do you currently find your partner?
(c) How physically attractive do other women currently find your partner?
(d) How sexually attractive do other women currently find your partner?

Take the average of these scores (i.e., add them up and divide by four) to find your partner's overall **attractiveness** score.

ANSWER

First the bad news: a recent study found that more attractive males make worse dads. Now the good news: this study was conducted on blue-black grassquits (a species of small bird). As far as I can tell, no similar study has yet been conducted on humans (by all means let me know if you find one).

The male grassquits' attractiveness was assessed by measuring the colour intensity (chroma) of their blue-black feathers. Their attentiveness as fathers was assessed by measuring the number of feeding visits that they made to the nest. And, yes, the prettier the male bird's feathers, the greater his inclination to leave parenting to the mother. Why? The 'aim' of the male grassquit – or, rather, of the genes for which he is merely an unwitting conduit – is to spread those genes by producing as many viable offspring as possible. From this perspective, feeding just one baby – particularly given that the mother will feed it anyway – is a waste of time that could be better spent playing away from home (grassquits are notionally monogamous). And the more attractive the male, the greater his opportunities for doing so. It is only the relatively dowdy male grassquits whose time is better spent caring for the gene-carriers that they have already fathered than fruitlessly seeking opportunities to father more.

Does the same cost-benefit analysis apply to humans? Who knows? But even if it doesn't now, it presumably applied to our ancestors. After all, if you go back far enough, we share a common ancestor with the grassquit, and the males of this long-lost species will have faced the same dilemma: father more kids, or be more of a father to the kids you already have?

Interestingly, the findings of one recent study suggest that, among humans, being a good or bad dad depends not on being good- or bad-looking but on how much the child in question looks like – and even smells like – you. The study was conducted with a rural farming community in Senegal. As is usual in this type of community, resources such as food are limited and are generally doled out by the father, who therefore has a strong incentive to favour the children that he is certain are his own: although polygamy is forbidden for women,* extramarital affairs can never be completely ruled out. Visual similarity was measured by showing raters photographs and asking them to pick the father of each child, from a choice of three (the more raters succeeded for each pair, the greater the similarity). Odour-similarity was measured in a similar way, but using T-shirts worn by children and fathers, rather than photographs. Mothers were then asked about the amount of time, emotional support and money that the fathers invested in each child. Sure enough, the level of investment in each child was predicted by both visual and odour-similarity to the father.

To end on a more heart-warming note, there is evidence that whether or not a man is good-looking, looking like a (potential) good dad is attractive in its own right. In the study in question, a male bumped into his 'sister' in a café and either cooed over, or ignored, her baby. Of course, the whole thing was a set-up for the purposes of a female stooge sitting at a nearby table, whom the male then approached for her phone number. Perhaps unsurprisingly, he succeeded far more often with women who had seen him interact with the baby. So, men, if you're looking to land a date, don't be a grassquit; instead show the object of your affections that you'd make a great dad.

---

* It is, however, allowed for men. That is, *polygyny* (where a man can have several wives) is permitted, but *polyandry* (where a woman can have several husbands) is not.

School of Collective Nouns

Given what we learned in the last section, I would like to propose a collective term for (males of) the species *Volatinia jacarina*: a *deadbeat* of grassquits. As we saw in Why Can't We Talk to the Animals? human languages are incredibly sophisticated in many ways, including – to take just one tiny example – the bewildering inventory of collective nouns that we have come up with to describe groups of other animals. We've all heard of a *flock* of birds, a *herd* of sheep, a *gaggle* of geese and perhaps even a *murder* of crows, but how many others do you know . . .?

An army of . . .
A convocation of . . .
A destruction of . . .
A flutter of . . .
An intrusion of . . .
A kindle of . . .
A pandemonium of . . .
A parliament of . . .
A plague of . . .
A scurry of . . .
A skulk of . . .
A stand of . . .
A tower of . . .
An unkindness of . . .
A wisdom of . . .

ANSWER

A wisdom of wombats
An unkindness of ravens
A tower of giraffes
A stand of flamingoes
A skulk of foxes
A scurry of squirrels
A plague of locusts
A parliament of owls
A pandemonium of parrots
A kindle of kittens (aww)
An intrusion of cockroaches
A flutter of butterflies
A destruction of cats (mostly used to refer to wild or feral cats)
A convocation of eagles
An army of ants

Who makes up these terms anyway? And how did some come to be seen as 'official' labels for groups of animals, while others – some, no doubt, no less apt – fell by the wayside. The English language has no official arbiter, but does, say, the *Académie Française* sit around debating what to call a group of crabs (it's a *panier* by the way, which presumably refers to the baskets that are used to catch them; in English, a *scuttle*)?

Who knows? But one thing is for sure: we humans aren't off the hook. There are literally hundreds of collective nouns for groups of people, including a *posse* of policemen, a *lechery* of priests, a *bloat* of programmers, an *anthology* of prostitutes, a *promise* of barmen, a *hack* of smokers, a *flash* of paparazzi and – one we can all relate to – an *embarrassment* of parents. As for me and my colleagues, we're either a *complex* of psychologists or (some of us) a *pomposity* of professors – ouch!

Looking back to the animal terms, what is interesting is that – aside from a handful that relate to purely physical properties or behaviours (e.g., a flutter of butterflies, a stand of flamingoes) –

they reveal our human tendency to anthropomorphise: ravens are mean, but owls and eagles are statesmanlike. Ants are disciplined, but parrots are chaotic. Wombats are wise (really?), but foxes are devious. Is it too much to suggest, then, that even something as apparently arbitrary as the words we use to refer to a group of creatures reflects our instinctive understanding that we humans are related to all other animals, and have the same basic needs and desires?

Signed, Sea-Lioned, Delivered

Probably, yes. Although evocative in their own way, the collective nouns used to refer to sea lions – *bob*, *crash* and *hurdle* – hardly suggest a deep understanding of sea-lion psychology. In fact, for these particular animals, the even-more-prosaic terms *group* or *band* are more appropriate. Why? Well, if you were looking to form an animal pop group or rock band, sea lions would be one of the best possible species: they can keep a beat.

But can you? To find out, we first need to get you tapping out a regular beat. (You'll also need a stopwatch: there's probably one on your phone.) Let's go for 120 beats per minute (bpm). There are two ways you can do this. The fun way is to choose a 120 bpm song and tap along. Aretha Franklin's *Respect*, with its driving beat, is the perfect choice, but you could also have The Pussycat Dolls' *Don't Cha*, Nirvana's *Come As You Are*, Bon Jovi's *It's My Life*, Sheryl Crow's *All I Wanna Do*, Katy Perry's *Teenage Dream*, Taio Cruz's *Dynamite*, or – if you must – Carly Rae Jepsen's *Call Me Maybe*.* If you don't have any of these songs to hand, you can do it the boring way: just count out two beats per second using your stopwatch.

Take as long as you need to get into the groove. Then, when you're confident that you know what 120bpm feels like, (re-)start your stopwatch and – without looking at it – count out 60 beats, trying to keep perfectly in time (it may help if you sing or hum your chosen song while you do so). As you tap the 60th beat, stop your stopwatch. If it shows 30 seconds, you kept the beat perfectly. If it shows less, you sped up. If it shows more, you slowed down. Most people speed up, though when I tried it just now, I slowed down, coming in at 31 seconds, probably because I was trying so hard not to speed up.

---

* Note to wedding DJs: that's quite a selection, and since they're all the same tempo, you can even mix them together.

Ronan, a Californian sea lion, was first trained to bob her head in time to a 120bpm metronome (or 'click track'), but rapidly learned to bob along to music, including Back Street Boys' *Everybody* (108bpm) and Earth, Wind and Fire's *Boogie Wonderland* (130bpm). Perhaps most impressively of all, Ronan was able to keep the beat on her own, when metronome clicks were removed (although she had it easier than you did, as only a single beat was dropped at any one time).

You might be thinking that this sounds a little – if you will excuse the pun – flippant. Why are psychologists, at the prestigious University of California no less, wasting their time training a sea lion to dance to the Back Street Boys? Actually, this finding is rather ground-breaking. We have long known that humans, cockatoos, budgerigars and parrots can keep a beat, while monkeys struggle. This led to the hypothesis that this ability is shown only by species that have evolved brain mechanisms dedicated to vocal learning and, in particular, vocal mimicry. (See Why Can't We Talk to the Animals? to read about a parrot whose mimicry led directly to his owner's divorce.) Since sea lions are not vocal mimics, Ronan's achievements give the lie to this hypothesis, and suggest instead that beat-keeping relies on more general timing mechanisms that are shared by many different animal species.

Excitingly, this keeps alive the long-standing dream of a band composed entirely of animals, as in – for example – the Grimm fairy tale of a donkey, dog, cat and rooster who head off to the Germany city of Bremen to form a band.

Of course, any animal band will need a singer . . .

Bird Is the Word (or Starlings in Their Eyes)

We all know that many birds are great at singing; but how good are you?

To find out, download the free karaoke software from www.mylittlekaraoke.com and see how you get on with *Jingle Bells*. If you score less than 5,000 (the current high score is around 10,000), then I officially declare you a worse singer than Groucho the parrot (see Web Links) who can belt out a pretty decent version of this festive favourite, along with *Campdown Races, Yankee Doodle Dandy, Alouette* and *How Much is that Doggy in the Window* (hey, nobody said parrots have good taste).

Groucho is a novelty, but real birdsong serves a serious purpose, or – more accurately – two serious purposes. The first is territorial. Most bird species are territorial, in that males will try to prevent other males coming on to their home turf, whether this is just a small area around the nest, a larger nesting and mating territory or an even larger all-purpose territory that is also used for courtship, feeding, foraging and so on. Songbirds generally defend the larger type of territory, using their songs to frighten off rivals. Experimental studies have shown that when male nightingales are played recordings of other male nightingales (but not other bird species), they increase the volume of their singing by as much as 5 decibels. So, as the comedian Russell Brand once pointed out, the nightingale whose 'plaintive anthem' inspired Keats's famous Ode was actually saying, 'Fuck off, this is my tree!' And it seems to work: a similar study with blackbirds found that, once the recorded singing got too loud,* the birds really did fuck off, and at

---

* Louder isn't always better, though. In fact, some species, such as song sparrows, always sing softly before an attack. Nobody is really quite sure why, but

great speed, ceding the territory to their (imaginary) rival before they got hurt. Perhaps most fascinating of all, a study of chipping sparrows found that males who are particularly good singers (defined, for this species, as those who produce faster trills) sometimes ventured on to the territory of a weaker-singing neighbour, in order to help him ward off a stronger-singing rival. It's unlikely that they were being altruistic, however: it is much better to have a weak-singing neighbour as a first line of defence than to allow a stronger singer – who may have one eye on your own territory – to move in next door (tellingly, sparrows never came to the aid of a neighbour who was a stronger singer than themselves).

The second purpose of birdsong (somewhat related to the first) is attracting females. In many species, males face off in singing competitions. The stakes are high: a study with supposedly monogamous black-capped chickadees found that females who witnessed their partner losing a sing-off (against a recording played by some mean experimenters) were more likely to 'cheat' on him, with the poor unsuspecting guy ending up a cuckold.

But, when trying to impress the ladies, just what counts as a 'better' song? The first rule of thumb is just 'the more the better'. In many species, including European starlings, blue tits and white-throated sparrows, females prefer males who simply sing for more hours per day. The rate at which males switch to a different song is also important: studies have shown that, for example, female pied flycatchers and zebra finches are impressed by males who pack more different songs into a five-minute period. Repertoire is important too. Females (including song sparrows and great tits) are not impressed by males who sing the same song over and over, and prefer those who have many different songs (or – for warbler species whose singing is continuous rather than divided into individual

---

one possibility is that these 'quiet threats' minimise the possibility of being overheard by both predators and other males (i.e., not the one that the singer is trying to frighten off but other rivals, perhaps ones who are bigger and stronger than the singer).

songs – many different 'syllables'*). While they appreciate variety in terms of repertoire, female birds are, well, xenophobic, preferring songs typical of the local population. Generally, a male singer has to be transported quite a long way before females will reject him as an outsider, but some species (including white-crowned sparrows and corn buntings) have local 'dialects' that apply to very small areas.†

Finally, we come to the more 'physical' properties of the song. As for territorial singing, both volume and trill rate matter: females like it fast and loud (and the latter applies also to the rather less musical mating calls of frogs, toads and insects). Females of some bird species, including swamp sparrows, also prefer songs that cover a wide pitch range (i.e., that include both very high and very low notes), particularly when this is combined with a high trill rate to produce a soaring or tumbling melody.

All this raises the question of why males' singing ability matters to females. Unlike the fancy eyespots on peacocks' feathers (see Spot the Difference), impressive singing seems to be not just an arbitrary ornamentation but a relatively direct indicator of males' relative fitness (in the Darwinian sense). It's not *only* that birds who can sing longer, faster, louder and so on are in prime physical condition, although that undoubtedly comes into it. Many experts – most notably, the biologist Stephen Nowicki – have argued that, because most of the brain structures necessary for singing develop when birds are young and vulnerable, an impressive singing ability demonstrates that the male in question not only managed to get through this difficult period unscathed but positively thrived.

So, we have a pretty good understanding of why birdsong is attractive to female birds, but why is it attractive to *us*? We humans love birdsong so much that we not only write odes to it but even

---

* A syllable is defined as a unit of any length that is repeated in exactly the same way; thus a syllable can be anything from a single note to a long sequence.

† Maybe this is just prejudice, but I'm sure we (humans) in the UK have far more of this than, say, Americans. The issue is complicated by the cross-cutting factor of social class, but even a single metropolitan area such as Merseyside or Greater Manchester has quite a few different local accents and dialects.

compose symphonies that mimic its patterns. Why don't we confer the same honour on cats' miaows, dogs' barks or whatever it is that elephants do?

The answer is that, as you probably already knew intuitively, birdsong is, completely objectively, *much* more like human music than are the various grunts, squeaks and squawks of these other species. Traditional Western scales are based around simple whole-number ratios. For example, a basic major chord consists of the first, third and fifth notes of the major scale (e.g., C, E and G). The frequency (or pitch) of a note is the number of times per second that a plucked string playing that note vibrates back and forth. Now what's special about a major chord (or the first, third and fifth notes of the scale) is that the ratio of the frequency of the highest note to the frequency of the lowest note is exactly 3:2, while the ratio of the middle note to the lowest note is exactly 5:4.* The reason that it feels natural to build scales out of perfect ratios is that these notes are indeed, 'natural', in the sense that they are already out there in the world: when a string is plucked, it gives off tones of several different frequencies at the same time. The loudest, the fundamental frequency, is what we normally think of as 'the pitch' of the note, but the string also gives off quieter harmonics.† These harmonics are perfect ratios of the fundamental frequency.

It was not until 2014 that it was demonstrated that the song of at least one bird species, the hermit thrush, is based around sim-

---

* Or, at least, it used to be. This tuning system (known as *just intonation*) used to be popular, but had all but died out by around 1800, replaced by the modern system of *equal temperament*. This system only *approximates* whole-number ratios, albeit pretty closely (e.g., the ratios of 1.5:1 [i.e., 3:2] and 1.25:1 [i.e., 5:4] have been replaced by ratios of 1.49:1 and 1.26:1). The advantage is that the equal temperament system allows songs in different keys to be played on the same keyed or fretted instrument without sounding out of tune. You can hear the difference on my web site: https://benambridge.wordpress.com/2014/04/24/the-secret-chord/

† This is why a single note on a guitar, violin or piano sounds rich and complex, while a synthesiser-produced 'pure tone', with no harmonics sounds thin and nasty.

ilar principles. A research team lead by a real-life Dr Doolittle (Dr Emily Doolittle, from the famously musical city of Seattle) used acoustic analysis to show that the frequencies of individual notes of the song tended to be whole-number multiples (i.e., harmonics) of the 'base frequency' (this note is not actually sung but corresponds to our notion of the 'key' of a song; the first note of our major scale). The result is a song that sounds remarkably human. But don't just take my word for it: follow the Web Links to listen to some examples provided by the researchers (the recordings have been slowed down so that both the speed and pitch range are more suitable for humans).

Completing the human-avian circle that we began with your (terrible!) singing, it turns out that birdsong researcher Dr Doolittle is herself an accomplished composer, and specialises in pieces inspired by our feathered friends. You can listen to her compositions by following the link below.

Web Links

See Emily play at: http://emilydoolittle.com/music/

Give Groucho marks (for his singing) at: https://www.youtube.com/watch?v=EZZ8FBxDZLg

Stringing You Along

Singing isn't all that birds offer in the way of entertainment. In fact, watching birds (particularly goldfinches*) draw water from a well used to pass for a good night in in the seventeenth century (well, nobody had invented Netflix) .

When animal psychologists first came on the scene a few hundred years later, they quickly realised that a string-pulling test is a quick and easy way to measure the intelligence of very different animal species. For example, in one common set-up two or more strings are laid out (sometimes bent, or even crossed), but only one is connected to a food reward. The question is simply how often the animal in question pulls the correct string. A review published in 2015 found that tests of this type had been conducted on 49 species of mammal (including apes, elephants, rats, cats, dogs, foxes and squirrels) and 46 species of bird (including pretty much any you can think of). Of course, this task is trivially simple for adults, but infants typically fail until around 16–20 months. So if you have a one-year-old handy, why not give them a try using the set-up shown here. The two outer strings should be longer than

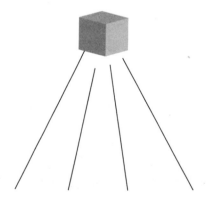

---

* The Dutch word for goldfinch is *putter*, whose literal translation is 'well-drawer'.

the two inner strings (say 35 and 30 cm respectively), so that the grabby ends of all four are roughly as easy for the child to reach. You'll need to have a few goes with a different string attached in order to check that they didn't just get lucky.

How did your little one get on? When this study was done under test conditions, children of this age were roughly equally split between success (pulling the correct string), failure (pulling at random) and intermediate performance (e.g., pulling two strings together, one of which was the correct one).

Why do some children fail at this apparently simple task? Until recently, most experts believed that one-year-olds do understand connectivity and how strings work (i.e., that the string has to be attached to the thing that you want) but failed at the task because of poor self-control: something for which infants are notorious. Even those who ultimately succeed at the task find it pretty difficult to resist the temptation to just pull *all* of the strings, in no particular order, just because the string-pulling 'game' is so much fun (ah, how quickly we lose that childish innocence!). However, a 2015 study found that children who succeeded at the string-pulling task themselves also glanced at the correct string in anticipation when watching an adult do the task. Children who failed at the task themselves did not. This suggests that, in fact, many 16–20-month-olds do not in fact understand the physics of string-pulling, putting them on a par with dogs, but behind most parrots.

The fact that we humans take more than a year to learn something that is almost second nature to parrots might not sound like great news for our species. But actually, the fact that humans are – compared to most other species – born relatively immature is one of the secrets of our success. Since a toddler's head can't fit through an adult pelvis (can you imagine?!), natural selection stumbled across the solution of having human heads continue to grow considerably after birth. Because a bigger brain means greater intelligence (see Are You Big-Headed?), this is one of the evolutionary adaptations that has allowed us *Homo sapiens* to outsmart not just all other human species (see Who Man Being?) but all of the species with whom we share our planet.

So don't be too concerned if – on the basis of this test, at least – your one-year-old is no smarter than a spaniel. If, on the other hand, he or she is string-pulling like a parrot, don't get too smug. Never mind the kids, it's time to ask . . .

Are You Smarter than an Orang-Utan?

We'll get to the titular chimpanzees in the next section, I promise. But, since they're pretty smart, I thought you should work up to it by first taking on an orang-utan (a much more distantly related great-ape species). Study the list of items below, then turn the page and try to list all seven (in any order):

Daffodil
Lizard
Fish
Apple
Seashell
Doctor
Rock

Are You Smarter than an Orang-Utan? Test

List the items in the box below

How did you do? Iris, an eleven-month-old female orang-utan tested at Smithsonian's National Zoo in Washington, DC, scored an average of just under 4/7. Of course, Iris didn't read the words on one page then write them on another; instead she completed both the learning and test part of the study using a touchscreen computer.

Now, 4/7 isn't a particularly impressive total, so you probably managed to beat it. But don't breathe a sigh of relief just yet; there's another way in which I'd like to compare your performance with Iris's. Despite her modest total, Iris showed evidence of something pretty clever: using a *strategy* to remember the list. You see, Iris completed this whole procedure not just once but several times, with the items in the training list in a different order for each run (like you, she could produce her own answers in any order).

If we split up each list that Iris produced into pairs (1st+2nd item; 2nd+3rd item; 3rd+4th item), a remarkable pattern emerges: roughly half of the pairs that Iris produced were the same every single time (e.g., *fish* always followed by *lizard*; *seashell* always followed by *rock*). This is remarkable, because it illustrates that Iris had spontaneously adopted a strategy of grouping together – for example – (a) animals and (b) types of stone. Of course, we can't tell exactly what grouping strategy she was using (animals vs moving things; stones vs hard things vs things you find at the beach). But the fact that she reliably produced at least some of the pairs in the same order each time demonstrates that she did indeed adopt some strategy or other, and stuck to it. This suggests that the brain mechanisms that enable us to develop strategies for learning and remembering are not exclusive to humans – or even humans and our closest relatives, chimpanzees and bonobos – but shared by all great apes.

Now, perhaps you've got such a great memory that you simply reeled off all seven items without needing a strategy. But if not, and it didn't occur to you to adopt one, then I'm afraid you've been outsmarted by an orang-utan. If this is the case, then you should probably feel a little apprehensive as we (finally) ask . . .

Are You Smarter than a Chimpanzee? #1

Along with bonobos (*Pan paniscus*), chimpanzees (*Pan troglodytes*) are our closest living relatives.* It is therefore unsurprising that these species are almost certainly the closest to humans in terms of intelligence (nobody would take this book seriously if its title were – say – *Are You Smarter than a Seal?*). But could it be that, in some ways, they're actually cleverer than us . . .?

Don't look yet, but below are two crosses made up of circles. Glance at them quickly and, without counting, guess which has more black circles:

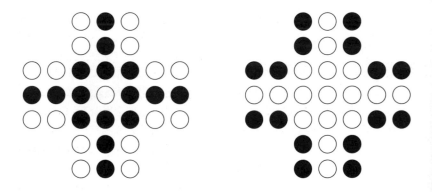

---

* Until very recently, chimps wore the crown as our closest living relative. However, since the 2012 publication of the bonobo genome sequence, the two species have had to share this distinction: each has almost exactly the same percentage overlap with humans. For some traits, chimpanzees and bonobos are more similar to humans than they are to each other.

ANSWER

Although both crosses contain 16 black circles (count 'em), most people think that the one on the left has more. The reason is that the black circles are grouped together in the left-hand cross, but spread out in the other. As a result, you see a good healthy dollop of black on the left, but just a smattering on the right. This finding is known as the *solitaire illusion*, as it is commonly illustrated using something like the arrangement of a solitaire game board (as in the version of the test you just took).

Chimpanzees completed a version of the study in which the black circles were replaced with blue M&Ms and the white circles with yellow cereal pieces. Their job was to select one of two boards, each displaying a pattern of the type that you saw on the previous page. Because chimpanzees much prefer M&Ms, they had a strong incentive to choose the board that contained – or *appeared* to contain – more of them.

Surprisingly, given that the illusion is a powerful one for humans, all of the chimpanzees tested (Lana, Panzee, Sherman and Mercury) chose each board equally often. It's not that they didn't understand the game. When the experimenters changed the setup so that one board really did contain more M&Ms than the other, the chimpanzees chose it without hesitating. So the fact that they picked at random when given the test you've just taken suggests that they simply don't fall for the illusion (they're certainly not capable of counting the individual M&Ms!).

So why do humans display the solitaire illusion when our (joint) closest animal cousins do not? The authors of the chimp study don't offer any suggestions, so here's one of my own: perhaps the illusion is a product of learning. Perhaps we humans have learned that when people have a small number of things they tend to spread them around for either aesthetic or practical reasons (e.g., candles on a cake, anchovies on a pizza, mobile phone masts in a country). On the other hand, when people have a large number of things (e.g., grains of rice, coffee beans) they tend to lump them together into one big mass or container, making them easy to handle. Perhaps, then, we have learned that spread-out normally means few and clumped-together normally means many. Because chimpanzees don't understand cakes or pizzas, let alone mobile phone masts or coffee jars, they've never learned the associations that trip us up when faced with this illusion.

Whether or not this explanation is correct, the fact that chimpanzees are impervious to an illusion that most of us humans fall for is certainly more than a little embarrassing for our species. At least in terms of certain kinds of visual-numerical intelligence, the answer to the question 'Are you smarter than a chimpanzee?' is, for most of us, 'um, no'.

Are You Smarter than a Chimpanzee? #2

Another way in which chimpanzees are particularly smart, and particularly humanlike, is in their ability to recognise each other's faces. Let's see how you compare. Don't look yet, but below are three faces, each made up from the top half of one face and the bottom half of another. Your job is to name the three celebrities who contributed the **top half** of each face, as quickly as possible. Ready? Go!

ANSWER

The celebrities who contributed the top half of each face were Will Smith, Brad Pitt and Halle Berry.

Here's another set of three composite faces. Again, your job is to name the three celebrities who contributed the **top half** of each face, as quickly as possible.

ANSWER

The celebrities who contributed the top half of each face were George Clooney, Madonna and David Beckham.

The celebrities who contributed the top half of each face were George Clooney, Madonna and David Beckham.

Did you find the second set of faces easier? That is, were you quicker to name all three? If so, you have just experienced what is known as the *composite face effect*. Most people find the first set relatively difficult because your instructions to identify the person contributing the top half of each face conflict with your face-processing mechanism, which is holistic in nature: you just can't help trying to process the whole face in one go. As a result, you find the chins of Morgan Freeman, Ryan Gosling and Gwyneth Paltrow – beautiful Hollywood chins though they are – distracting and confusing. In the second set of pictures, the bottom half of the face is offset, meaning that you are more easily able to bypass your usual holistic face-processing mechanism and to focus on the top half. As a result, you find the chins of Robert De Niro, Katy Perry and Tom Cruise relatively easy to ignore (probably the most statistically improbable sentence I've ever written).

Chimpanzees, of course, are not asked to identify celebrity chimps (who would they have used – those ones from the PG Tips adverts? Nim Chimpsky?). Instead, they complete a *match-to-sample* task of the type we met in *Pigeon's Treat* (*A Taxi Attacks*). They are shown the face of a sample chimpanzee (effectively the 'celebrity') and must pick the face whose top half matches it (from a choice of two). Just like humans, chimpanzees do better on this task when the top and bottom halves of the composite faces are offset, rather than aligned to create a confusing face that is, nevertheless, difficult to ignore. This finding shows that the dedicated holistic face-processing mechanism that we humans enjoy is not specific to our species but was already present in the common ancestor that we share with chimpanzees.

Are You Smarter than a Chimpanzee? #3

Chimpanzees' smartness doesn't just stop at recognising faces; they are also good at recognising and interpreting the expressions that those faces are pulling. So, how do you compare? Look at the five chimpanzee expressions below, and see if you can guess which emotion is being expressed in each case.

ANSWER

Well, are you? Let's find out.

The *bared-teeth* display is usually interpreted as something like a human smile, perhaps particularly the polite 'pleased to meet you' smile that we give when being introduced to a new person. This expression is most often used when approaching, hugging or initiating play with another individual, or to defuse an aggressive approach. It has obvious similarities with a human smile, although if we humans try to mimic the expression exactly, it looks a bit more like a false smile or grimace (see below).

You could be forgiven for failing to get the second one, since it has no real human equivalent (see how the human equivalent below is pretty uninterpretable). This chimpanzee is making a *pant-hoot*: a call that signifies excitement, particularly in response to food.

The third expression is pretty straightforward. This is a *play face*: an expression that tends to be used only during play (and hence is used mainly by young chimps). The adult human pulling the equivalent expression below doesn't particularly look like he's playing, but is at least attempting to convey a kind of relaxed happiness (if not entirely convincingly!).

You should also have little difficulty with the fourth: a *scream*. Just like its human equivalent below, it indicates fear, distress, aggression or conflict. Many experts believe that chimpanzees have distinct submissive and aggressive screams, although it's probably not possible to tell them apart in a photo.

The final expression is a *bulging-lip* display, analogous to a human angry face. Grrrr!

Chimpanzees are, of course, adept at recognising and responding appropriately to these expressions when they are made in the wild (if they weren't, there would be no point in making them). But scientists have also come up with some ingenious ways of testing chimps' ability to recognise facial expressions in the lab. In one type of study, chimpanzees are trained to use a joystick to move a cursor on a computer screen. They are then shown a picture of one expression (e.g., bared-teeth 'smile') and can win a food reward by dragging the cursor to a picture of a different chimpanzee making the same expression, as in the illustration below.

Studies using this method have shown that chimps are capable of correctly categorising all of the major expression types discussed above, even when they are made by individuals they have never met, and have seen only in a photograph.

And most impressively: they can use a joystick! I'm old enough to remember when all computer games were controlled in this way, and it was not at all unusual to encounter humans who simply could not get the hang of it. So if you want to know whether you're really smarter than a chimpanzee, why not see if you can beat this one at Ms Pacman?

Web Links
Chimp: https://www.youtube.com/watch?v=r7ttRaXlnfs
You: http://www.mspacman1.com/

It's Been Emotional

So, it seems that we share at least some of the basic emotions of chimpanzees, but what about other, more complex, emotions? Do animals share these, and – if so – which animals? Just primates? Mammals? *All* animals? For that matter, are emotions universal across all humans (as Darwin thought), or are there some that are specific to particular cultures? And why do we have emotions in the first place? To start to get a handle of some of these difficult questions, let's complete a widely used emotion questionnaire.

Please recall a recent situation in which you strongly experienced each of the seven emotions shown below. For each situation please tick off the symptoms and reactions shown in the table.

Symptom/reaction	Joy	Fear	Anger	Sadness	Disgust	Shame	Guilt
lump in throat							
change in breathing							
stomach troubles							
feeling cold/shivering							
feeling warm/pleasant							
feeling hot/cheeks burning							
heart beating faster							
muscles tensing/trembling							
perspiring/moist hands							
laughing/smiling							
crying/sobbing							
screaming/yelling							
gesturing							
moving towards people/things							
withdrawing from people/things							
aggression							
silence							

ANSWER

In the original version of this study, a similar questionnaire was given to participants from thirty-seven different countries, spanning all five continents. The results were remarkably similar. The table below shows boxes that were ticked by around half of all participants across the globe, so, wherever you're from, it's very likely that your pattern of ticks corresponds closely to that shown (just for stats geeks, the actual percentages of people ticking each box are shown too).

Symptom/reaction	Joy	Fear	Anger	Sadness	Disgust	Shame	Guilt
lump in throat	14%	29%	25%	56% ✓	21%	23%	28%
change in breathing	20%	47% ✓	37%	24%	18%	21%	17%
stomach troubles	3%	21%	11%	19%	21%	11%	15%
feeling cold/shivering	2%	36%	8%	22%	14%	11%	12%
feeling warm/pleasant	63% ✓	1%	1%	1%	1%	2%	1%
feeling hot/cheeks burning	18%	14%	32%	9%	11%	40%	18%
heart beating faster	40%	65% ✓	50% ✓	26%	23%	35%	27%
muscles tensing/trembling	10%	52% ✓	43%	27%	25%	23%	22%
perspiring/moist hands	11%	37%	21%	16%	13%	26%	17%
laughing/smiling	85% ✓	4%	3%	4%	4%	14%	4%
crying/sobbing	9%	15%	15%	55% ✓	7%	9%	13%
screaming/yelling	8%	13%	22%	6%	7%	3%	4%
gesturing	20%	17%	27%	14%	18%	18%	17%
moving towards people/things	39%	14%	10%	13%	5%	5%	8%
withdrawing from people/things	1%	18%	18%	30%	22%	27%	24%
aggression	0%	5%	29%	6%	11%	4%	6%
silence	13%	51% ✓	24%	61% ✓	39%	47% ✓	48% ✓

What makes the similarities even more remarkable is that participants were asked, in each case, to think about a *specific recent situation* in which they experienced that emotion. Because not every symptom/reaction occurs every time (e.g., not every sad event causes a lump in the throat), the cross-cultural similarities would almost certainly have been even higher, had people been asked if each emotion *ever* leads to the relevant symptom/reaction. As the authors of this study acknowledge, the fact that emotional responses are universal does not demonstrate that they are genetically hard-wired in all humans, but universality is a necessary first step (i.e., if emotional responses had turned out *not* to be universal, then they could not be genetic).

So, emotions seem to be pretty universal across humans, but how about animals?* We have to be careful here. On the one hand, if animals show behaviours or physical responses that we would label as fear, anger, sadness and so on in humans, it seems unfair to deny animals these emotions *solely* because they're animals. And they certainly seem to. For example, rats (and cats) show an increase in heart rate and blood pressure when showing 'fear' or aggression ('anger'?), but a decrease when they must 'cope' with prolonged pain ('sadness'?, 'resignation'?). They also sigh with 'relief' (i.e., when they hear a tone indicating that an expected electric shock will not actually be delivered). On the other hand, we must be careful to avoid anthropomorphism (inappropriately ascribing human characteristics to non-human things). Even plants make 'choices' (e.g., where to grow) and have 'desires' that they will go to some lengths to fulfil (e.g., growing through concrete to reach sunlight), but most of us would agree that they don't have emotions.

---

* Some people argue that emotions, by definition, require consciousness (e.g., 'sadness' is only 'sadness' because we can reflect on and talk about it), and that consciousness, by definition, is exclusive to humans (it's whatever we have that animals lack; see *Epilogue*). This is a perfectly valid point of view, but clearly one that we must set aside if the question of whether or not animals have emotions is to be meaningful, rather than just an argument over our definitions.

How can we find some middle ground? Biologist Marian Dawkins* invites us to consider the case of a rat in a maze who learns that a left turn results in food, and a right turn in an electric shock (or vice versa). Unlike plants growing towards sunlight, the response (turning left rather than right, or vice versa) is totally arbitrary, and so can't be genetically hard-wired. She argues, then, that the only way to explain the rat's ability is to assume that it has some punishment-reward system which allows it to associate certain behaviours with 'feeling better' and others with 'feeling worse'. In other words, it seems that we are pretty much forced to assume that the humble rat – and, indeed, any animal that can do this type of simple learning – has *something* like the emotions 'pleasure' and 'suffering' (we'll get into a more detailed discussion of animal 'pain' in Every Body Hurts?).

Why do we have emotions? The evolutionary advantage of seeking out 'pleasure' (e.g., food and sex) and avoiding 'suffering' (e.g., injury and starvation) is obvious, but what about the other emotions? Enter Robert Plutchik and his 'psychoevolutionary' theory of emotions, which, he holds, are shared by all animals, even those without a nervous system, at least 'in some form'. Although it may seem odd to attribute 'anger' to an amoeba or 'surprise' to a slug, Plutchik's argument is that the most important aspect of the emotion is its associated behaviour and adaptive evolutionary function (anger→attack→destroy obstacle; surprise→stop→gain time for orientation), rather than the touchy-feely subjective experience (which is what we humans tend to focus on). Plutchik's eight emotions, behaviours and adaptive functions are listed below.

---

* If you were wondering, then, yes, she is married to Richard; both are Oxford professors.

Stimulus	Subjective experience	Behaviour	Adaptive function
Food, sex etc.	Joy	Keep hold of it; do it again	Gain/conserve resources
Member of one's group/family	Trust	Groom	Gain support
Threat	Fear	Run away	Avoid injury
Unexpected event	Surprise	Stop	Gain time for orientation
Lose food, partner etc.	Sadness	Cry	Get it back
Unpleasant taste/smell	Disgust	Vomit	Get rid of poison
Obstacle or enemy	Anger	Attack	Remove obstacle or enemy
New object or territory	Anticipation	Explore	Gain knowledge

We could add more of our own. For example, perhaps 'shame' and 'guilt' (for instance, in response to upsetting a friend or partner) prompt us to avoid repeating the behaviour, and so have the adaptive function of preventing us from losing support, food or sex. And presumably 'jealousy' – sometimes called the useless emotion – prompts us to take steps to keep rivals away from our mating partners (which, far from being useless, would be highly adaptive from an evolutionary perspective). The problem, as with many evolutionary explanations, is that while this all sounds perfectly plausible, it is going to be pretty much impossible to confirm or deny. The philosopher Jerry Fodor points out that it is easy to come up with a semi-plausible evolutionary 'just-so story' for just about any human behaviour, emotion or desire.

So, while Plutchik may be taking it a bit far with his claim that *all* animals have at least eight major emotions, there does seem to be good reason to think that many emotions are not unique to humans, or even primates, but are shared by many other mammals.

Psycho Gorilla, Qu'est-ce que c'est?

As we saw in the previous section, most emotions seem to be pretty universal among humans. One category of possible exceptions, however, is psychopaths. Psychopaths seem to lack, or at least show very low levels of, one of the most widespread and adaptive emotions: fear. As a result, they are supremely self-confident, particularly in social situations, remain calm under pressure and are unfazed by new and potentially dangerous situations. This fearlessness is one of the three key aspects of the psychopathic personality. We'll meet the other two in a moment, but first let's take a psychopath test . . .

. . . for chimpanzees. That's right, a group of researchers based in Georgia, USA, decided to investigate psychopathy in chimpanzees. Whatever for? Well, if our closest evolutionary ancestors* show psychopathy too, this is good evidence that the disorder has at least some genetic and/or biological basis. By comparing chimpanzees and humans, we can investigate the relative contributions of these 'nature' factors and the 'nurture' factors that also seem to contribute to psychopathy in humans (for example, there is evidence that childhood deprivation also plays a role). The researchers' aim was therefore to come up with a test that could be taken by both chimpanzees and humans (and presumably other great apes too – hence the title of this section). You may, of course, have to use a bit of artistic licence when applying the scale to yourself. For example, where Q1 asks about ability to 'threaten or take food from other chimpanzees', you should substitute 'threaten or take things from other people'.

---

* Well, joint closest with bonobos; see Are You Smarter than a Chimpanzee? #1. The titular gorillas are third.

		Least like me				Most like me		
1. **Dominant.** Is able to displace, threaten or take food from other chimpanzees. May express high status by decisively intervening in social interactions.	*mL, mS*	1	2	3	4	5	6	7
2. **Dependent.** Often relies on other chimpanzees for leadership, reassurance, touching, embracing and other forms of social support.		7	6	5 *mL*	4 *mS*	3	2	1
3. **Anxious.** Hesitant, indecisive, tentative, jittery.		7	6	5 *mL / mS*	4	3	2	1
4. **Fearful.** Reacts excessively to real or imagined threats by displaying behaviours such as screaming, grimacing, running away or other signs of anxiety or distress.		7	6 *ML*	5 *mS*	4	3	2	1
5. **Bold.** Daring, not restrained or tentative. Not timid, shy or coy.		1	2	3 *mS*	4	5	6 *mL*	7
6. **Timid.** Lacks confidence, is easily alarmed and is hesitant to venture into new social or non-social situations.		7	6	5 *mS / mL*	4	3	2	1
7. **Impulsive.** Often displays some spontaneous or sudden behaviour that could not have been anticipated.		1	2 *mL*	3	4 *mS*	5	6	7
8. **Inventive/spontaneous.** More likely than others to engage in novel behaviours: e.g., using new devices or materials in their enclosure.		1	2	3 *mS*	4 *mL*	5	6	7
9. **Irritable.** Often seems in a bad mood or is impatient and easily provoked to anger or exasperation and consequent antagonistic behaviour.		1	2	3 *mS*	4	5 *mL*	6	7
10. **Excitable.** Easily aroused to an emotional state. Becomes highly aroused by situations that would cause less arousal in most chimpanzees.		1	2	3 *mL*	4	5	6	7 *mS*

	Least like me				Most like me		
11. **Calm.** Equable, restful. Reacts to others in an even, calm way. Is not easily disturbed or agitated.	7	6	5 *mS*	4	3	2 *mL*	1
12. **Socially inept/intrusive.** Acts inappropriately in a social setting.	1 *mL mS*	2	3	4	5	6	7
13. **Jealous/attention-seeking.** Often troubled by others who are in a desirable or advantageous situation such as having food, a choice location or access to social groups. May attempt to disrupt activities or make noise to get attention.	1	2 *mS*	3	4 *mL*	5	6	7
14. **Kind/considerate.** Often consoles others in distress to provide reassurance.	7	6	5	4	3	2 *mS mL*	1
15. **Affectionate/friendly.** Seems to have a warm attachment or closeness with other chimpanzees. This may entail frequently grooming, touching, embracing or lying next to others.	7	6	5	4	3	2 *mS mL*	1
16. **Bullying.** Overbearing and intimidating towards younger or lower-ranking chimpanzees.	1 *mL mS*	2	3	4	5	6	7
17. **Manipulative.** Is adept at forming social relationships for its own advantage, especially using alliances and friendships to increase its social standing. Chimpanzee seems able and willing to use others.	1 *mL mS*	2	3	4	5	6	7
18. **Stingy.** Is excessively desirous or covetous of food, favoured locations or other resources in enclosure. Is unwilling to share these resources with others.	1 *mS*	2	3	4 *mL*	5	6	7

mS 7
mL (1)

ANSWERS

Questions 1–6 measure **boldness**, the fearlessness that is characteristic of psychopathy. If you scored a total of 6 to 12, you are not at all bold, but timid. If you scored 13 to 35, you are averagely bold. If you scored 36 or more, you are very bold.

Questions 7–13 measure **disinhibition**. Because of the constraints imposed by polite society, most of us are quite inhibited. We try to act appropriately in social settings, rather than drawing attention to ourselves by becoming over-excited or irritated, or by acting on our worst impulses. Psychopaths don't hold themselves back in this way. They show not inhibition, but disinhibition. If you scored a total of 7 to 14, you are not at all disinhibited, but inhibited. If you scored 15 to 41, you are averagely (dis)inhibited. If you scored 42 or more, you are very disinhibited.

Questions 14–18 measure **meanness**. Most of use try to be kind, considerate, affectionate and generous. Again, psychopaths don't bother. If you scored a total of 5 to 10, you are not at all mean, but kind. If you scored 11 to 29, you are averagely kind/mean. If you scored 30 or more, you are very mean.

A true psychopath would score very highly for all three traits: boldness, disinhibition and meanness. In fact, he (most are male) would probably score 7 on pretty much every question. But he almost certainly wouldn't admit to being a psychopath. Psychopaths tend to reject the idea that they have a problem, and justify being – for example – bold, bullying and stingy by saying things like, 'Well if you don't stand up for yourself and take what you want, others will ride roughshod over you'. If you think this sounds like something that a CEO would say, you're right! Psychopathy is a sliding scale, not all or nothing, and CEO is the profession that, on average, scores highest on psychopathy scales, followed by lawyers, salesmen and surgeons (care workers, nurses and therapists score lowest).*

---

* You can read more about these 'professional psychopaths' in my previous book, *Psy-Q*.

Back to the chimps, and the findings of this study suggest that the so-called *triarchic* account of psychopathy (boldness + disin-hibition + meanness) fits this species just about as well as it fits humans. Chimps' scores on the questionnaire were able to predict their performance in tasks such as dealing with an apparent in-truder (a human mannequin) and delaying gratification (see Let's Get Rat-Arsed). So, if, next time you're at the zoo, you're worried that the chimpanzee has a mean-looking glint in his eye, don't be afraid to whip out this questionnaire. And if he scores highly on all three measures of psychopathy, take Talking Heads' advice and run, run, run, run, run, run, run away.

Every Body Hurts?

Psychopaths, be they human or chimpanzee, are almost completely insensitive to the pain of other creatures. The rest of us, on the whole, do care and generally try our best to avoid causing pain to others, whether human or not. But do *all* animals feel pain?

Historically, we humans have tended to assume that other animals don't feel pain (or, at least, not in the same way that we do). For example, the famous French philosopher René Descartes argued that pain requires consciousness, which is unique to humans (a debate we will soon explore; see Self-Awareness: Have Monkeys Cottoned On?). Nowadays, although we are no closer to understanding what pain might *feel like* for other species, few scientists would endorse the view that pain is unique to humans. But how far should we take this? Is pain restricted to just the higher mammals, or does it apply to literally every body?

What's your hunch? In the table below, circle 'Yes' or 'No' for the organisms that you think can and can't feel pain respectively.

Organism	Hurts ... sometimes?	Organism	Hurts ... sometimes?
Horse	Yes / No	Zebrafish	Yes / No
Rabbit	Yes / No	Squid	Yes / No
Rat	Yes / No	Prawn	Yes / No
Parrot	Yes / No	Crab	Yes / No
Chicken	Yes / No	Bee	Yes / No
Frog	Yes / No	Fruit fly	Yes / No
Tortoise	Yes / No	Bacteria	Yes / No
Trout	Yes / No		

ANSWERS

First we have to clarify exactly what we mean by 'feels pain'. Researchers working in this field make a distinction between *pain* and *nocioception*. Nocioception is the ability to move away from something that is potentially damaging. Even bacteria are capable of nocioception, but nobody seriously thinks that they feel pain, least of all because they lack a nervous system. So we wouldn't want to say that any animal that tries to avoid potential damage 'feels pain'. Although we can't, of course, read animals' minds, we need *some* evidence that they not only try to escape damaging stimuli but also experience some kind of unpleasant feeling if they fail to do so.

Fortunately, Dr Lynne Sneddon (one of my colleagues at the University of Liverpool) has helpfully put together a checklist of just the type of evidence we'd need. Although there is no single 'smoking gun', if we can tick off all of these points (there are seventeen of them!) for a particular animal, we can be pretty confident that they do indeed feel pain.

- **Wired for pain.** First of all, you have to (1) have **nocioceptors**, cells that are capable of detecting pain and (2) the ability to **move away** from harmful stimuli (both of which rule out plants but include bacteria and all animals). But it's not enough just to *have* nocioceptors: they have to be linked to both (3) the **central nervous system** and (4) the **brain**, particularly areas that are related to 'fear' and learning (otherwise there's no way you could experience any kind of subjective feeling of pain).
- **Ouch, that hurt!** Next, you have to show (5) **physiological responses,** such as an increase in adrenaline, heart rate, blood pressure, body temperature and breathing rate. You also have to show (6) **behavioural responses** (other than simply moving away), such as pulling a face (rats, rabbits and horses all do this), crying, yelping or rearing up and (7) **altered behavioural choices or preferences** (e.g., Marian Dawkins's rat which learns to turn left for food rather than right for a shock, or vice versa).

Other key behaviours are (8) **protecting** or **avoiding the use of a damaged limb** (e.g., limping) and (9) **rubbing or licking your wounds** (crabs do both). If you have 'pain on the brain', you will often (10) **act preoccupied**: for example, failing to flee from predators (fish do this).

- **Watch out!** Animals (including crabs) will (11) take great steps to avoid pain, but also (12) show 'trade-off' effects with things that they like (e.g., starving fish will swim into a part of the tank that contains food, even if they know they will get an electric shock). They also (13) show 'pain relief learning', learning to love a signal (e.g., a buzzer) which shows that an expected shock will not actually be delivered.

- **Pass the paracetamol.*** The flip side of pain is analgesia, or pain relief. If animals respond to, and seem to care about, painkillers, then presumably they can experience pain (else, why would they bother?). To pass this test, an animal must (14) have brain cells that respond to pain-relieving drugs (e.g., frogs have opiod **receptors**), (15) **self-administer** these drugs given the chance (chickens do this), even if (16) they must **pay some cost** to do so (injured zebrafish will swim into a tank that they otherwise hate if there are painkillers in the water). Perhaps the clincher (17) is if, when the animal finally gets the painkilling drugs, all the physiological and behavioural responses listed above **decrease** or stop altogether. For example, parrots pull their foot away from a flame, but can stand the heat longer if they've been given painkillers.†

Although not every species has been tested against every criterion, the only definite fails are insects, which don't (8) protect their wounds or (10) seem to be driven to distraction by pain (as we saw in *Insex*, an earwig doesn't let a little thing like a snapped-

---

* Q: Why are there no aspirin in the jungle? A: Parrots eat 'em all.

† Sadly, given the previous footnote, this was not paracetamol but tramadol (a cheap heroin alternative beloved of Scottish junkies, as name-checked in the nihilistic comedy series *Frankie Boyle's Tramadol Nights*.)

off penis spoil the mood). Other than that, pretty much all animals pass all of the tests (i.e., you should have ticked 'Yes' for all the organisms in the list, except for bacteria). So if you meet a meat-lover, factory farmer, hunter or pescatarian who confidently tells you that such-and-such an animal can't feel pain, feel free to show them this chapter. It seems Michael Stipe & Co. were right all along: *Every* Body Hurts.

Oh, Beehave

One thing that really hurts is getting stung. But where is the worst place to get stung, and which insect has the most painful sting? Thanks to two intrepid scientists, Michael Smith and Justin ('Schmidt! That Hurt!') Schmidt, we now have the answers to both of these questions. Smith applied bee stings to his body at the twenty-five different locations shown below. Which do you think hurt the least . . . and the most?

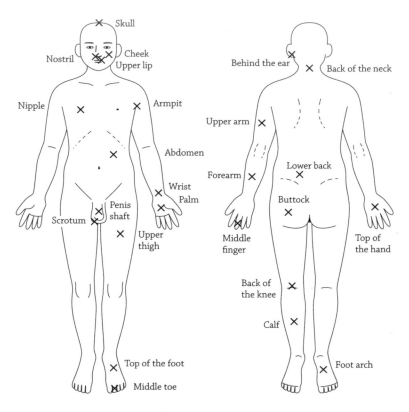

Schmidt stung himself with no fewer than seventy-eight different species and gave each a rating on a four-point scale (along with some vivid descriptions). Arrange the insects listed below from least painful to most.

Fire ant

Yellowjacket (a North American wasp)

Sweat bee

Bullet ant

Bullhorn acacia ant

Red harvester ant

Tarantula hawk (not a hawk but a terrifying wasp that hunts tarantulas in order to paralyse them and lay its eggs inside)

ANSWER

According to Smith (and I don't see anyone queuing up to verify his findings), the most painful places to be stung by a bee are the nostril, the upper lip and – unsurprisingly – the penis and scrotum. The least painful places are the skull, the tip of the middle toe, the upper arm and – just slightly more painful – the buttocks.

And here are Schmidt's ratings (from least to most painful), along with descriptions (from the June 2008 edition of *The Buzzword*, the magazine of the West Sound Beekeepers' Association).

Sweat bee (1/4): Light, ephemeral, almost fruity. A tiny spark has singed a single hair on your arm.

Fire ant (1.2/4): Sharp, sudden, mildly alarming. Like walking across a shag carpet and reaching for the light switch.

Bullhorn acacia ant (1.8/4): A rare, piercing, elevated sort of pain. Someone has fired a staple into your cheek.

Yellowjacket (2/4): Hot and smoky, almost irreverent. Imagine W. C. Fields extinguishing a cigar on your tongue.

Red harvester ant (3/4): Bold and unrelenting. Somebody is using a drill to excavate your ingrown toenail.

Tarantula hawk (4/4): Blinding, fierce, shockingly electric. A running hairdryer has been dropped into your bubble bath.

Bullet ant (off the chart): Pure, intense, brilliant pain. Like fire-walking over flaming charcoal with a three-inch rusty nail in your heel.

Ouch!

Taking the P***

Another thing that's pretty painful is when you really – I mean *really* – need the loo. In fact, if you're one of those heroic readers who tries to finish books in one sitting, you're probably feeling pretty uncomfortable right now, particularly now I've mentioned it (now *that's* psychology). Go on then, run along. Oh, just one thing if you wouldn't mind: can you time yourself? Not how long you take to get going, but how long it lasts from start to finish.

ANSWER

Was it twenty-one seconds? A recent IgNobel-prizewinning study found that all mammals weighing more than 3kg – from cats, dogs and goats to horses, lions and elephants – take almost exactly the same time to empty their bladder.* Yes, in the time that it takes a 3½kg cat to dispose of 5ml of urine (that's just 1/66th of a standard Coke can) a 3½ *thousand* kg elephant unleashes a positively torrential 18 litres (that's 54 cans' worth). So, as you can imagine, it comes out at quite a pace. How did the authors find out? The answer is simple: 'We filmed the urination of 16 animals and obtained 28 videos of urination from YouTube' – a sentence I never thought I'd read in the *Proceedings of the National Academy of Sciences*. Generously, the authors have made these videos available on their web site. So if you would like to see them – purely in the interests of science, of course – follow the Web Link below.

What about smaller animals? It turns out that they share neither the universal urination duration nor the satisfying gushing action of their larger cousins. Animals under 3kg (rats, bats, mice and so on) get the whole business done in under two seconds – for some, just a *hundredth* of a second – in a disappointing dribble of tiny droplets.

For some reason, the authors didn't include humans in their study (perhaps YouTube frowns on videos of human urination), but don't worry; if you haven't already done so, I'm sure you have all of the necessary equipment to carry out your own investigation.

Web Link
See animals pee at: http://www.pnas.org/content/111/33/11932.full

---

* Little did the authors of this study know that they had already been scooped by the UK Garage act So Solid Crew, who in 2001 released *21 Seconds To Go*.

Self-Awareness: Have Monkeys Cottoned On?

Suitably relieved, we are now pretty much ready to wrap up this book, one of the themes of which has been that – counter to the mantra of human uniqueness that we have drummed into ourselves over the centuries – we humans are really not so different from other animals after all. But before revisiting this and our other main themes in the final section, we need to consider perhaps the biggest possible counter-example: *self-awareness*. This is something that human children figure out pretty early on, as you can see for yourself, if you happen to have one lying around (perhaps the one you tested for Stringing You Along?).

Sit the child on your lap and read her a story. When you are confident that she is thoroughly engrossed, surreptitiously put a sticker on her forehead (or use a soft pen to draw a dot). If she reaches for her forehead, or otherwise seems to notice, give up and try again another time. If you got away with it, wait at least one minute, then show the child her reflection in a mirror. If she now reaches for her forehead (and not the reflection of it in the mirror), then she has passed the test, and is officially *self-aware*. Exactly what this means is a tricky and much-debated issue, but a child who passes this test has managed to form at least some kind of link between the 'me' that she experiences from within and the 'me' that is visible to the outside world. Certainly, the test measures self-knowledge in some meaningful way, as only children who have passed this test seem to show 'self-conscious' emotions such as pride or embarrassment.*

---

* That said, there seem to be different levels of self-awareness. The ability to point to oneself in a picture when asked 'Where are you?' does not generally emerge until the child is at least two, or even older if she is wearing a mask in

So, do other animals pass the test? And what does it mean if they do? Well, in another blow for human uniqueness, it has been known since the early 1970s that chimpanzees pass (to make absolutely sure they can't feel the dot going on, they are given a full anaesthetic beforehand). As for other primates, the other great apes – bonobos, orang-utans and (perhaps) gorillas – pass, while monkeys don't, suggesting that self-awareness is a recent evolutionary development. So it was extremely big news when, back in 1995, Harvard professor Marc Hauser published a paper reporting that cotton-top tamarins – a species of monkey – passed the test when their white 'cotton-tops' were dyed pink, blue, purple or green (the little punks).

Alas, if it sounds too good to be true, it probably is. Several researchers, including Gordon G. Gallup Jr, who conducted the original chimpanzee study, raised various objections regarding Hauser's methods. This prompted Hauser to conduct a new version of the study, published in 2001, in which the tamarins failed to pass the test. These findings still stand, but Hauser retracted an article on a different cotton-top tamarin study after a Harvard investigation found that 'The results of the experiment were therefore knowingly and falsely reported by Prof. Hauser'. Hauser himself, in the words of a report by the Office of Research Integrity (ORI), 'neither admits nor denies committing research misconduct but accepts ORI has found evidence of research misconduct'. Gallup, meanwhile, raised a few eyebrows of his own with the 2002 publication of a study entitled *Does Semen Have Antidepressant Properties?**

Subsequently, a handful of non-primate species have passed the mirror test, including dolphins, elephants and – perhaps –

---

the picture. These tests measure children's ability to understand that the self is permanent (i.e., 'a picture taken a few days ago is still me') and not dependent on appearance (i.e., 'that doesn't look like me, but because I remember wearing that mask, I know that the person behind the mask is me').

* I'm not suggesting any research misconduct on the part of Gallup; the raised eyebrows are solely to do with the subject-matter.

magpies.* This is still big news, of course, but less so than if the cotton-top tamarins had passed. Dolphins, elephants and magpies – unlike tamarins – are too distantly related to great apes for the capacity of self-awareness to have begun with a common ancestor. Instead this is an example of *convergent evolution*: two or more unrelated (or extremely distantly related) species developing a similar trait, presumably because it offers a similar advantage to each.

Just what might this advantage be? Gallup suggests that self-awareness is the key to understanding the mental states of others. Self-awareness involves understanding that 'me' consists of both internal thoughts and desires, on the one hand, and an outwardly observable body, which performs various actions, on the other. From here it's not such a mental leap to work out that, since other members of one's species have similar bodies and perform similar actions, they probably have similar internal thoughts and desires: a similar 'me'. In this way, self-awareness could be the source of socially intelligent behaviours such as deception, pretending, empathy and gratitude. As we have already seen, chimpanzees do seem to engage in at least some of these behaviours, such as deception (see A Fishy Tale). If there is a difference between humans and chimpanzees, it is that we tend to use our understanding of others' mental states for co-operation (see Why Can't We Talk to the Animals?), whereas they tend to use it for competition (as in the Fishy Tale study, where the chimpanzee uses deception to ensure that he – and not the human competitor – gets the banana).

So, does this mean that chimpanzees (and bonobos, orangutans, dolphins and elephants) have *consciousness*? . . .

---

* A subsequent study with jackdaws raised the possibility that birds might actually be responding to the feeling of the sticker on their feathers.

Epilogue: Relative Values Revisited

The weird and wacky phenomena that we have encountered on our caper through the animal kingdom may at times seem bizarre, but they all start to make sense when seen through a certain lens: Darwin's theory of evolution. Throughout this book we've seen evolution happening before our own eyes. In fact, as we saw with the silver-fox and anole-lizard experiments (discussed in A Walking Dogtionary and Who Man Being?) we can not only *see* evolution happening, but even *make* it happen. And as if this weren't enough, we can quite literally spell out, letter by letter, the DNA changes that occur as one species evolves into another (see Who Man Being? and Are You Smarter than a Chimpanzee? #1).

Once we accept that each and every creature in the animal kingdom is a relative, the idea that the difference between humans and other animals is relative rather than absolute starts to sound less radical, and more like common sense. We've seen that both humans and other animals ...

... when it comes to intelligence and cognition ... inherit intelligence, learn routes, track the probabilities of events in our environment, fall for mathematical fallacies, downplay options that we have previously rejected, are risk-averse and risk-seeking when faced with gains and losses respectively, learn words, understand how word order changes meaning, use swarm intelligence, count objects (using a ready-reckoning system for larger quantities), add up, memorise the location of food (even when it has been hidden by others), use mnemonics to learn lists, follow logical reasoning (even in highly abstract scenarios based on relative rather than absolute properties), use string and tools, fall for visual illusions and recognise ourselves in a mirror;

... when it comes to personality and emotions ... differ on personality traits such as extraversion and agreeableness, have a preference for one hand/paw over another (and show associated personality characteristics), can be psychopaths, feel pain and display emotions (and their physical correlates);

... **when it comes to sex** ... use looks (including symmetry) to choose a mate with good genes, sigh to release sexual tension, develop sexual fetishes, are (just the males) spurred on by sperm-competition scenarios and try to minimise the possibility of cuckoldry;

... **drugs and rock 'n' roll** ... benefit from psychological and pharmaceutical treatments for anxiety, enjoy taking drugs (but show impairments when under the influence), keep a beat and sing;

... **when it comes to competition and sociability** ... recognise friends and foes, deceive rivals when competing for food, seek out social interaction, use camouflage, forage cost-effectively for food, establish a pecking order, process faces holistically and interpret others' facial expressions.

Now it could be that all of this is just one big coincidence. Alternatively, and – to my mind – more plausibly, it could be that whether human, hyena or housefly, these traits are directly or indirectly crucial to the long term survival of your genes. Of course, nobody is denying that humans are better – in most cases, *way* better – than other animals at these tasks and abilities. The point is simply, to quote Darwin again, that the difference is 'one of degree and not of kind'.

Before we accept this conclusion, however, we need to explore the one final frontier of uniqueness that we humans seem desperate not to cross: *consciousness*. Just what is consciousness? The idea is notoriously difficult to define, but goes way beyond mere *self-awareness* (as discussed in the previous section). Consciousness *requires* self-awareness, of course, but what most people mean by the term is something more: that vague feeling we all have that 'This is me. Here I am. I can not only see stuff and do stuff, but even think stuff; like this stuff I'm thinking right now.' Isn't consciousness unique to humans, something that 'we' have and 'they' don't?

Logically, there are four possibilities, which we'll explore in turn. We can skip quickly over the first – that animals have consciousness and humans don't – on the basis that, to my knowledge, this is an argument that nobody has ever made. But only if we can reject the second possibility – that both humans and other animals have consciousness – and the third – that neither group does – do we have any grounds for accepting the fourth: that consciousness indeed represents the pinnacle of human uniqueness.

Perhaps surprisingly, given that the idea of human uniqueness has gained so much currency in the wider world, the dominant view among scientists who actually know about this stuff is that both humans and other animals have consciousness. In 2012 a summit of all the bigwigs in the field of neuroscience led to the publication of the Cambridge Declaration on Consciousness:

Convergent evidence indicates that non-human animals have the neuroanatomical, neurochemical, and neurophysiological substrates of conscious states along with the capacity to exhibit intentional behaviours. Consequently, the weight of evidence indicates that humans are not unique in possessing the neurological substrates that generate consciousness. Non-human animals, including all mammals and birds, and many other creatures, including octopuses, also possess these neurological substrates.

What this means in plain English is that all of the structures and circuits in the human brain that are involved in conscious feelings – which can be boosted or reduced by artificially stimulating or suppressing these structures – are also present in some form in (at least) mammals and birds. What is more, if these structures are artificially boosted or suppressed in non-human animals (i.e., by 'zapping' or 'freezing' the relevant part of the brain), they show the behaviours that one would expect if they were experiencing the same conscious feelings (e.g., excitement, fear, sadness). Of course, we have no way of knowing what excitement subjectively *feels like* to a mouse. But given that it has the necessary brain circuits and acts excited, it would seem churlish to deny the mouse this conscious experience. Indeed, perhaps the only reason to do so would be if we had decided in advance that animals cannot experience conscious feelings by definition, whatever the neuroscience tells us.

That's all very well, but you may have spotted a bit of sleight of hand in the above, in that I started talking about conscious *feelings* such as excitement, fear and sadness, rather than consciousness per se. These feelings are certainly *part of* what it is like to be conscious, but – for most people – consciousness means something more: the background hum of 'this is me, here I am' that we experience even in the absence of excitement, fear, sadness or any particular feeling. What about this type of consciousness?

Here we turn to Princeton University Professor of Neuroscience, Michael Graziano. Graziano invites us to consider a military commander with models representing his armies arranged on a map (like the board game Risk). The models are crude and short on detail – in fact, they barely look like the armies that they represent at all but they're good enough to get the job done. Like a general controlling his armies, the brain must control its attention: you can't look at, listen to or think about two things at once. So say you're looking at a cup on your desk, or thinking about a film you saw yesterday. Your brain can't store a complete, highly detailed representation of the cup or the film, but it can store a model of each that's good enough for controlling its attention (*cup . . . film . . . . cup . . film . . . . cup . . . . film*). Suppose that the brain now turns its attention to itself (*cup . . . film . . . brain*). Just as for the cup and the film, it's not possible to store a complete, highly detailed representation; instead we need a model. That imperfect model of the brain – what the brain sees when it focuses its attention on itself – that, according to Graziano, is consciousness. And given that even the humblest animals have attention that they can focus on things in the world, including their own brains, they too presumably experience a form of consciousness.

Graziano's theory is difficult to get a handle on, and – by his own admission – not particularly satisfying. We seem to want consciousness to be some magical essence, rather than something so prosaic as the attention system's model of the brain. But an advantage of this theory is that – unlike the 'magic essence' view –

it explains what consciousness is for: controlling one's attention isn't spiritual or magical, but it's as useful as hell ('A predator!!').

We now turn to the third possibility: that neither humans nor animals have consciousness. The idea – as espoused most famously by the philosopher Daniel Dennett – is that consciousness (in its 'this is me, here I am' incarnation) is essentially a trick that the brain plays on itself.

What does it mean for the brain to play a trick on itself? Here's an example. Look at the Bernardo Bellotto painting below:

Dresden from the Right Bank of the Elbe above the Augustus Bridge

If we zoom in on the bridge a bit, we can see lots of people . . .

... or so it seems.

If we zoom in a bit more it turns out that they are carefully placed blobs of paint that merely *suggest* people. Let's let Dennett take up the argument in his own words:

Now, what does your brain do when it takes the suggestion? [Does it] ... send little painters out to fill in all the details in your brain somewhere? I don't think so. Not a chance. But then, how on Earth is it done? ... The brain just makes you think that it's got the detail there. You think the detail's there, but it isn't there. The brain isn't actually putting the detail in your head at all. It's just making you expect the detail.

According to Dennett, consciousness is just the same: you think it's there, but it isn't. There's no magical consciousness – your brain is just tricking you into expecting it.

If this seems far-fetched, bear in mind that we have been here many times before. We used to believe that all living things have an *élan vital*, or life force, that somehow courses through them. We used to believe that forces such as magnetism and gravity are transmitted through *the ether*, an invisible substance that fills all the otherwise-empty space. Further back, we used to believe in a *homunculus*, a little man inside each of us that takes in information from the eyes and ears and decides how to respond. We have since learned to live without these concepts; they now strike us as laughable. Yet, at the time, these things seemed as if they just *had to be* there. But they never were; our brains were just filling in the detail that they expected. Consciousness, according to Dennett, is just another idea that we will have to learn to live without.

At first glance, the Cambridge–Graziano view – that all species have consciousness – and the Dennet view – that none does – seem like polar opposites. But, deep down, both are really saying the same thing: all of the phenomena that we can observe in others – be they human or animal – can be explained by common-or-garden neuroscience – the brain systems responsible for hearing, memory, vision, self-awareness, feelings, emotions, attention and so on – and beyond that there is really nothing left to explain.

This brings us to the fourth and final possibility, that – beyond all this – humans have a subjective internal experience of consciousness that demands an explanation. For all the efforts of the neuroscientist and philosophers, this remains the position that – if push came to shove – most people would plump for. If you find yourself in this category, let me try to give your unshakable conviction just the tiniest of tremors with another of Dennett's experiments: this time a thought experiment.

A *zimbo* is a special type of zombie. Not the flesh-eating kind, quite the opposite: the whole point about zimbos is that, in terms of outward appearance – which includes anything that they say or do – they resemble normal human beings. In fact, they *are* normal human beings with just one difference: they lack consciousness. Now, let's say that your partner or best friend has been accused of being a zimbo. Of course, to say that you are curious to find out whether or not this is true would be a gross understatement.

But what would you do? Remember that, to all outward appearances, zimbos are exactly like humans. This means that, for any question you can ask, a zimbo can give a perfectly reasonable human-like answer. So if a zimbo is to talk convincingly about feeling sad, it has to have not only a 100 per cent realistic simulation of human sadness, but a 100 per cent realistic simulation of thinking about sadness, of thinking about thinking about sadness, and so on ad infinitum. But if a zimbo has all of these things, what exactly differentiates its simulations from the real thing? Or, to put it another way, how do you know that *you're* not a zimbo? It certainly *seems* to you as though you have consciousness, but then if you are to be capable of carrying on a conversation with real humans about consciousness – which, by definition, all zimbos are – then you would require a simulation of consciousness that is 100 per cent indistinguishable from the real thing. Even to you.

I therefore put it to you that – along with birds, mammals, octopuses and so on – you *are* a zimbo: a creature with non-miraculous brain structures that support feelings and experiences, including the experience of being an experiencer, with no mystical cherry of consciousness on top.

All this is necessarily pretty speculative, but this may not be the case for ever. Suppose that, at some point in the future, our understanding of neuroscience reaches such a point that it is possible to upload a simulation of – say – a bat's brain into our own. Then we really could experience what it's like to be a bat,* including any subjective experience of consciousness, or lack thereof.

In conclusion, perhaps future research will discover that there really is some qualitative difference between us and other animals. In the meantime, history – in the form of the *élans*, ethers and homunculi now consigned to its dustbin – tells us that it is probably wise to avoid positing some magical essence in lieu of an explanation.

Ah yes, history. If there really is an important difference between humans and other animals, the best candidate is not biological but cultural. If we are unique in any way, it is in our ability – born of our capacity for co-operation – to pass knowledge not only between individuals but also between generations. I hope that, by exploring the nature of the relationship between humans and other animals, this book has contributed, in some small way, to this process of cultural evolution: to our *shared* human consciousness.

---

* As readers in the know will have immediately realised, the reason I chose a bat is that perhaps the most famous philosophy paper of all time, Thomas Nagel's 'What is it like to be a bat?', concluded that we can never know.

References

Introduction: Relative Values
https://www.bible.com/bible/1/gen.1
http://www.firstthings.com/web-exclusives/2015/07/more-than-in-gods-image
http://www.gutenberg.org/files/15707/15707-h/15707-h.htm
http://www.gutenberg.org/ebooks/7116
http://www.gutenberg.org/ebooks/2009

An Expensive Cappuccino
Catapano, R., Buttrick, N., Widness, J., Goldstein, R., and Santos, L. R. (2014). Capuchin monkeys do not show human-like pricing effects. *Frontiers in Psychology*, 5, 1330. doi:10.3389/fpsyg.2014.01330

Shiv, B., Carmon, Z., and Ariely, D. (2005). Placebo effects of marketing actions: consumers may get what they pay for. *Journal of Marketing Research*, 42, 383–93 10.1509/jmkr.2005.42.4.383

Waber, R. L., Shiv, B., and Carmon, Z. (2008). Commercial features of placebo and therapeutic. *Journal of the American Medical Association* 299, 1016–17. 10.1001/jama.299.9.1016

The Eye of the Beeholder
Ambridge, B. (2014). *Psy-Q*. London: Profile, pp .207–10

Frey, F. M., and Bukoski, M. (2014). Floral symmetry is associated with flower size and pollen production but not insect visitation rates in Geranium robertianum (Geraniaceae). *Plant Species Biology*, 29(3), 272–80

Rodríguez, I., Gumbert, A., de Ibarra, N. H., Kunze, J., and Giurfa, M. (2004). Symmetry is in the eye of the 'beeholder': innate preference for bilateral symmetry in flower-naïve bumblebees. *Naturwissenschaften*, 91(8), 374–7

Shepherd, K., and Bar, M. (2011). Preference for symmetry: only on Mars? *Perception*, 40(10), 1254

The Usual Waspects
Baracchi, D., Petrocelli, I., Chittka, L., Ricciardi, G., and Turillazzi, S. (2015). Speed and accuracy in nest-mate recognition: a hover wasp prioritises face recognition over colony odour cues to minimise intrusion by outsiders.

Proceedings of the Royal Society of London B: Biological Sciences, 282(1802), 20142750

de Souza, A.R., Alberto Mourão Júnior, C., Santos do Nascimento, F., and Lino-Neto, J. (2014). Sexy faces in a male paper wasp. *PLoS ONE* 9(5): e98172. doi:10.1371/journal.pone.0098172

Gherardi, F., Aquiloni, L., andand Tricarico, E. (2012). Revisiting social recognition systems in invertebrates. *Animal Cognition*, 15(5), 745–62

Spot the Difference

Dakin, R., and Montgomerie, R. (2011). Peahens prefer peacocks displaying more eyespots, but rarely. *Animal Behaviour*, 82(1), 21–8

Dakin, R., and Montgomerie, R. (2013). Eye for an eyespot: how iridescent plumage ocelli influence peacock mating success. *Behavioral Ecology*, 24(5), 1048–57.doi: 10.1093/beheco/art045

How the Giraffe Got His Neck

Agaba, M., Ishengoma, E., Miller, W. C., McGrath, B. C., Hudson, C. N., Reina, O. C. B., . . . and Praul, C. A. (2016). Giraffe genome sequence reveals clues to its unique morphology and physiology. *Nature Communications*, 7, 1–8

Cameron, E. Z., and du Toit, J. T. (2007). Winning by a neck: tall giraffes avoid competing with shorter browsers. *The American Naturalist*, 169(1), 130–35

Heard, E., and Martienssen, R. A. (2014). Transgenerational epigenetic inheritance: myths and mechanisms. *Cell*, 157(1), 95–109

Mitchell, G., Van Sittert, S. J., and Skinner, J. D. (2009). Sexual selection is not the origin of long necks in giraffes. *Journal of Zoology*, 278(4), 281–6

Simmons, R. E., and Scheepers, L. (1996). Winning by a neck: sexual selection in the evolution of giraffe. *American Naturalist*, 771–86

http://www.realclearscience.com/blog/2014/03/end_the_hype_over_epigenetics__lamarckian_evolution.html

A Fishy Tale

Bshary, R., and Grutter, A. S. (2005). Punishment and partner switching cause cooperative behaviour in a cleaning mutualism. *Biology Letters*, 1(4), 396–9

Hare, B., Call, J., and Tomasello, M. (2006). Chimpanzees deceive a human competitor by hiding. *Cognition*, 101(3), 495–514

Fish Are Amazing

Galea, L. A., and Kimura, D. (1993). Sex differences in route-learning. *Personality and Individual Differences*, 14(1), 53–65

Silverman, I., Choi, J., Mackewn, A., Fisher, M., Moro, J., and Olshansky, E. (2000). Evolved mechanisms underlying wayfinding: further studies on the hunter–gatherer theory of spatial sex differences. *Evolution and Human Behavior*, 21(3), 201–13

Smith, C., Philips, A., and Reichard, M. (forthcoming). Cognitive ability is heritable and predicts the success of an alternative mating tactic. *Proceedings of the Royal Society of London B: Biological Sciences*, 282(1809), 20151046

Trzaskowski, M., Yang, J., Visscher, P. M., and Plomin, R. (2014). DNA evidence for strong genetic stability and increasing heritability of intelligence from age 7 to 12. *Molecular psychiatry*, 19(3), 380–84

Are You Big-Headed?

Aiello, L. C., and Wheeler, P. (1995). The expensive-tissue hypothesis: the brain and the digestive system in human and primate evolution. *Current Anthropology*, 36(2), 199–221

Bushby, K. M., Cole, T., Matthews, J. N., and Goodship, J. A. (1992). Centiles for adult head circumference. *Archives of Disease in Childhood*, 67(10), 1286–7

Jerison, H. (2000) The evolution of intelligence. In Sternberg, R.J. (ed.), *Handbook of Intelligence*. New York: Cambridge University Press, pp. 216–44

Kotrschal, A., Rogell, B., Bundsen, A., Svensson, B., Zajitschek, S., Brännström, I., ... and Kolm, N. (2013). Artificial selection on relative brain size in the guppy reveals costs and benefits of evolving a larger brain. *Current Biology*, 23(2), 168–71

McDaniel, M. A. (2005). Big-brained people are smarter: a meta-analysis of the relationship between in vivo brain volume and intelligence. *Intelligence*, 33(4), 337–46

Rushton, J. P., and Ankney, C. D. (2009). Whole brain size and general mental ability: a review. *International Journal of Neuroscience*, 119(5), 692–732

Sternberg, R. J. (2012). Intelligence. *Wiley Interdisciplinary Reviews: Cognitive Science*, 3(5), 501–11

Goatbusters

Herbranson, W. T., and Schroeder, J. (2010). Are birds smarter than mathematicians? Pigeons (*Columba livia*) perform optimally on a version of the Monty Hall Dilemma. *Journal of Comparative Psychology*, 124(1), 1–13

Pigeon's Treat

Klein, E. D., Bhatt, R. S., and Zentall, T. R. (2005). Contrast and the justification of effort. *Psychonomic Bulletin & Review*, 12(2), 335–9

White, K. G., and Magalhães, P. (2015). The sunk cost effect in pigeons and people: a case of within-trials contrast? *Behavioural Processes*, 112, 22–8

Zentall, T. R. (2015). When animals misbehave: analogs of human biases and suboptimal choice. *Behavioural Processes*, 112, 3–13

Zentall, T. R., and Clement, T. S. (2002). Memory mechanisms in pigeons: evidence of base-rate neglect. *Journal of Experimental Psychology: Animal Behavior Processes*, 28(1), 111

MMMMonkeying Around

Customer: What sandwiches have you got?
Waiter: Chicken salad and roast beef.
Customer: OK, I'll have the roast beef please.
Waiter: Oh yes, we also have tuna.
Customer: Right. In that case, I'll have the chicken salad.

Brehm, J. W. (1956). Postdecision changes in the desirability of alternatives. *The Journal of Abnormal and Social Psychology*, 52(3), 384

Egan, L. C., Santos, L. R., and Bloom, P. (2007). The origins of cognitive dissonance: evidence from children and monkeys. *Psychological Science*, 18(11), 978–83

Frank, R. H. (2008). *The Economic Naturalist: Why Economics Explains Almost Everything*. New York: Random House

More Monkeyconomics

Kahneman, D. (2011). *Thinking, Fast and Slow*. New York: Macmillan

Lakshminarayanan, V. R., Chen, M. K., and Santos, L. R. (2011). The evolution of decision-making under risk: Framing effects in monkey risk preferences. *Journal of Experimental Social Psychology*, 47(3), 689–93

Ape, Man United

Attrill, M. J., Gresty, K. A., Hill, R. A., and Barton, R. A. (2008). Red shirt colour is associated with long-term team success in English football. *Journal of Sports Sciences*, 26(6), 577–82

Greenlees, I. A., Eynon, M., and Thelwell, R. C. (2013). Color of soccer goalkeepers' uniforms influences the outcome of penalty kicks. *Perceptual & Motor Skills*, 117(1), 1–10

Morris, D. (1981). *The Soccer Tribe*. London: Jonathan Cape

http://www.thebeautiful92.com/about

A Stable Personality?

Gosling, S. D., and John, O. P. (1999). Personality dimensions in non-human animals: a cross-species review. *Current Directions in Psychological Science*, 8, 69–75

Gosling, S. D., Kwan, V. S. Y., and John, O. P. (2003). A dog's got personality: a cross-species comparative approach to evaluating personality judgments. *Journal of Personality and Social Psychology*, 85, 1161–9

http://www.sheknows.com/pets-and-animals/articles/1084150/funny-dog-names

The Truth about Cats and Dogs
Gosling, S. D., Sandy, C. J., and Potter, J. (2010). Personalities of self-identified 'dog people' and 'cat people'. *Anthrozoös*, 23, 213–22

Dog-Person or Person-Dog?
Augusti, E. M., Melinder, A., and Gredebäck, G. (2010). Look who's talking: pre-verbal infants' perception of face-to-face and back-to-back social interactions. *Frontiers in Psychology*, 161(1), 1–7

Kujala, M. V., Kujala, J., Carlson, S., and Hari, R. (2012). Dog experts' brains distinguish socially relevant body postures similarly in dogs and humans. *PloS one*, 7(6), e39145 (images taken from this article reproduced under Creative Commons Licence)

Törnqvist, H., Somppi, S., Koskela, A., Krause, C. M., Vainio, O., amd Kujala, M. V. (2015). Comparison of dogs and humans in visual scanning of social interaction. *Royal Society Open Science*, 2(9), 150341

http://www.sciencemag.org/news/2015/04/feature-solving-mystery-dog-domestication

A Walking Dogtionary: Test
Kaminski, J., Call, J., and Fischer, J. (2004). Word learning in a domestic dog: evidence for 'fast mapping'. *Science*, 304(5677), 1682–3

Pilley, J. W., and Reid, A. K. (2011). Border collie comprehends object names as verbal referents. *Behavioural Processes*, 86(2), 184–95

Who Man Being?
Currat, M. and Excoffier, L. (2011) Strong reproductive isolation between humans and Neanderthals inferred from observed patterns of introgression. *Proceedings of the National Academy of Sciences of the United States of America*, 108, 15129–34

Krings, M., Stone, A., Schmitz, R. W., Krainitzki, H., Stoneking, M., and Pääbo, S. (1997). Neandertal DNA sequences and the origin of modern humans. *Cell*, 90(1), 19–30

Lau, A. N., Peng, L., Goto, H., Chemnick, L., Ryder, O. A., and Makova, K. D. (2009). Horse domestication and conservation genetics of Przewalski's horse inferred from sex chromosomal and autosomal sequences. *Molecular Biology and Evolution*, 26(1), 199–208

Neves, A. G. M. and Serva, M. (2012) Extremely rare interbreeding events can explain Neanderthal DNA in living humans. *PLoS ONE*, 7(10): e47076. doi:10.1371/journal.pone.0047076

Noonan, J. P. (2010). Neanderthal genomics and the evolution of modern humans. *Genome Research*, 20(5), 547–53

Short, R. V., Chandley, A. C., Jones, R. C., and Allen, W. R. (1974). Meiosis in interspecific equine hybrids. *Cytogenetic and Genome Research*, 13(5), 465–78

Stuart, Y. E., Campbell, T. S., Hohenlohe, P. A., Reynolds, R. G., Revell, L. J., and Losos, J. B. (2014). Rapid evolution of a native species following invasion by a congener. *Science*, 346(6208), 463-6

A Formidable Problem

Alhanjouri, M., and Alfarra, B. (2011). Ant colony versus genetic algorithm based on travelling salesman problem. *International Journal of Computer Technology and Applications*, 2(3), 570–78

Dorigo, M., and Gambardella, L. M. (1997). Ant colony system: a cooperative learning approach to the traveling salesman problem. *Evolutionary Computation, IEEE Transactions on*, 1(1), 53–66

Schmickl T, Karsai I (2014) Sting, carry and stock: how corpse availability can regulate de-centralized task allocation in a ponerine ant colony. *PLoS ONE* 9(12), e114611. doi:10.1371/journal.pone.0114611

The Tower of Han(t)oi

Reid, C. R., Sumpter, D. J., and Beekman, M. (2011). Optimisation in a natural system: Argentine ants solve the Towers of Hanoi. *Journal of Experimental Biology*, 214(1), 50–58

Losing Your Marbles

Grünbaum, D. (1998). Schooling as a strategy for taxis in a noisy environment. *Evolutionary Ecology*, 12(5), 503–22

Krause, J., Ruxton, G. D., and Krause, S. (2010). Swarm intelligence in animals and humans. *Trends in Ecology and Evolution*, 25(1), 28–34

Seeley, T. D., Visscher, P. K., and Passino, K. M. (2006). Group decision making in honey bee swarms. *American Scientist*, 94(3), 220

Wallis, K. F. (2014). Revisiting Francis Galton's forecasting competition. *Statistical Science*, 29(3), 420–24

Circle of Life

de Fockert, J., Davidoff, J., Fagot, J., Parron, C., and Goldstein, J. (2007). More accurate size contrast judgments in the Ebbinghaus Illusion by a remote culture. *Journal of Experimental Psychology: Human Perception and Performance*, 33(3), 738

Doherty, M. J., Campbell, N. M., Tsuji, H., and Phillips, W. A. (2010). The Ebbinghaus illusion deceives adults but not young children. *Developmental Science*, 13(5), 714–21

Salva, O. R., Rugani, R., Cavazzana, A., Regolin, L., and Vallortigara, G. (2013). Perception of the Ebbinghaus illusion in four-day-old domestic chicks (*Gallus gallus*). *Animal Cognition*, 16(6), 895–906

Pige-lusi-on # 1

Skinner, B. F. (1948). 'Superstition' in the pigeon. *Journal of Experimental Psychology*, 38(2), 168–72

Wiseman, R., and Watt, C. (2004). Measuring superstitious belief: why lucky charms matter. *Personality and Individual Differences*, 37(8), 1533–41

Pige-lusi-on # 2

Nakamura, N., Watanabe, S., and Fujita, K. (2009). Further analysis of perception of the standard Müller-Lyer figures in pigeons (*Columba livia*) and humans (*Homo sapiens*): effects of length of brackets. *Journal of Comparative Psychology*, 123(3), 287–94

Mathemalex

Pepperberg, I. M. (2006). Grey parrot (*Psittacus erithacus*) numerical abilities: addition and further experiments on a zero-like concept. *Journal of Comparative Psychology*, 120(1), 1–11

Pepperberg, I. M., and Gordon, J. D. (2005). Number comprehension by a grey parrot (Psittacus erithacus), including a zero-like concept. *Journal of Comparative Psychology*, 119(2), 197–209

http://www.scientificamerican.com/article/what-is-the-origin-of-zer/

A Shaggy Dog Story

Sheldrake, R., and Smart, P. (2000). A dog that seems to know when his owner is coming home: videotaped experiments and observations. *Journal of Scientific Exploration*, 14, 233–55

Wiseman, R., Smith, M., and Milton, J. (1998). Can animals detect when their owners are returning home? An experimental test of the 'psychic pet' phenomenon. *British Journal of Psychology*, 89, 453–62

Dogtanian and the Three Must-Get Pairs

Duranton, C., Rödel, H. G., Bedossa, T., and Belkhir, S. (2015). Inverse sex effects on performance of domestic dogs (*Canis familiaris*) in a repeated problem-solving task. *Journal of Comparative Psychology*, 129(1), 84–7

Lejbak, L., Vrbancic, M., and Crossley, M. (2009). The female advantage in object location memory is robust to verbalisability and mode of presentation of test stimuli. *Brain and Cognition*, 69, 148–53

Turning Japanese

Herman, L. M., Richards, D. G., and Wolz, J. P. (1984). Comprehension of sentences by bottlenosed dolphins. *Cognition*, 16(2), 129–219

Kako, E. (1999). Elements of syntax in the systems of three language-trained animals. *Animal Learning and Behavior*, 27(1), 1–14

Why Can't We Talk to the Animals?

http://www.freerepublic.com/focus/f-chat/1548521/posts

Brownell, C., Svetlova, M., and Nichols, S. (2009). To share or not to share: when do toddlers respond to another's needs? *Infancy*, 14(1), 117–30

Cheney, D., and Seyfarth, R. (1990). Attending to behaviour versus attending to knowledge: examining monkeys' attribution of mental states. *Animal Behaviour*, 40(4), 742–53

Jensen, K., Hare, B., Call, J., and Tomasello, M. (2006). What's in it for me? Self-regard precludes altruism and spite in chimpanzees. *Proceedings of the Royal Society of London B: Biological Sciences*, 273(1589), 1013–21

Silk, J. B., Brosnan, S. F., Vonk, J., Henrich, J., Povinelli, D. J., Richardson, A. S., . . . and Schapiro, S. J. (2005). Chimpanzees are indifferent to the welfare of unrelated group members. *Nature*, 437(7063), 1357–9

Ueno, A. and Matsuzawa, T. (2004). Food transfer between chimpanzee mothers and their infants. *Primates* 45, 231- 9

Warneken, F. and Tomasello, M. (2009). Varieties of altruism in children and chimpanzees. *Trends in Cognitive Science*, 13, 397–402

Something To Crow About?

DeWall, C. N., Baumeister, R. F., and Masicampo, E. J. (2008). Evidence that logical reasoning depends on conscious processing. *Consciousness and Cognition*, 17(3), 628–45

Jelbert, S. A., Taylor, A. H., and Gray, R. D. (2015). Reasoning by exclusion in New Caledonian crows (*Corvus moneduloides*) cannot be explained by avoidance of empty containers. *Journal of Comparative Psychology*, 129(3), 283–90

Verzoni, K., and Swan, K. (1995). On the nature and development of conditional reasoning in early adolescence. *Applied Cognitive Psychology*, 9(3), 213–34

Students vs Squirrels, Sorta

Wills, A. J., Lea, S. E., Leaver, L. A., Osthaus, B., Ryan, C. M., Suret, M. B., ... and Millar, L. (2009). A comparative analysis of the categorisation of multidimensional stimuli: I. Unidimensional classification does not necessarily imply analytic processing; evidence from pigeons (*Columba livia*), squirrels (*Sciurus carolinensis*), and humans (*Homo sapiens*). *Journal of Comparative Psychology*, 123(4), 391–405

Box Clever

Leighty, K. A., Grand, A. P., Pittman Courte, V. L., Maloney, M. A., and Bettinger, T. L. (2013). Relational responding by eastern box turtles (*Terrapene carolina*) in a series of color discrimination tasks. *Journal of Comparative Psychology*, 127(3), 256–64

Is There A Dog-tor in the House?

Anestis, M. D., Anestis, J. C., Zawilinski, L. L., Hopkins, T. A., and Lilienfeld, S. O. (2014). Equine-related treatments for mental disorders lack empirical support: a systematic review of empirical investigations. *Journal of Clinical Psychology*, 70(12), 1115–32

Kamioka, H., Okada, S., Tsutani, K., Park, H., Okuizumi, H., Handa, S., ... and Honda, T. (2014). Effectiveness of animal-assisted therapy: a systematic review of randomised controlled trials. *Complementary Therapies in Medicine*, 22(2), 371–90

Karagiannis, C. I., Burman, O. H., and Mills, D. S. (2015). Dogs with separation-related problems show a 'less pessimistic' cognitive bias during treatment with fluoxetine (Reconcile™) and a behaviour modification plan. *BMC Veterinary Research*, 11(1), 80

McCulloch, M., Jezierski, T., Broffman, M., Hubbard, A., Turner, K., and Janecki, T. (2006). Diagnostic accuracy of canine scent detection in early-and late-stage lung and breast cancers. *Integrative Cancer Therapies*, 5(1), 30–39

Macpherson, K., and Roberts, W. A. (2006). Do dogs (*Canis familiaris*) seek help in an emergency? *Journal of Comparative Psychology*, 120(2), 113–19
https://www.psychologytoday.com/blog/animals-and-us/201411/does-animal-assisted-therapy-really-work

Let's Get Rat-Arsed

Bickel, W. K., Yi, R., Landes, R. D., Hill, P. F., and Baxter, C. (2011). Remember the future: working memory training decreases delay discounting among stimulant addicts. *Biological Psychiatry*, 69(3), 260–65

Coffey, S. F., Gudleski, G. D., Saladin, M. E., and Brady, K. T. (2003). Impulsivity and rapid discounting of delayed hypothetical rewards in cocaine-dependent individuals. *Experimental and Clinical Psychopharmacology*, 11(1), 18–25

Perry, J. L., Larson, E. B., German, J. P., Madden, G. J., and Carroll, M. E. (2005). Impulsivity (delay discounting) as a predictor of acquisition of IV cocaine self-administration in female rats. *Psychopharmacology*, 178(2-3), 193–201

Poulos, C.X., Le, A.D., and Parker, J.L. (1995) Impulsivity predicts individual susceptibility to high levels of alcohol self-administration. *Behavioral Pharmacology* 6, 810–14

Oh, What a Tangled Web We Weave

Deussing, J. M. (2007). Animal models of depression. *Drug Discovery Today: Disease Models*, 3(4), 375–83

Meldrum, M. (1998). 'A calculated risk': the Salk polio vaccine field trials of 1954. *British Medical Journal*, 317(7167), 1233

Witt, P. N. (1971). Drugs alter web-building of spiders: a review and evaluation. *Behavioral Science*, 16(1), 98–113

Insex

Baer, B., Morgan, E. D., and Schmid-Hempel, P. (2001). A nonspecific fatty acid within the bumblebee mating plug prevents females from remating. *Proceedings of the National Academy of Sciences*, 98(7), 3926–8

Hotzy, C., Polak, M., Rönn, J. L., and Arnqvist, G. (2012). Phenotypic engineering unveils the function of genital morphology. *Current Biology*, 22(23), 2258–61

Kamimura, Y., and Matsuo, Y. (2001). A 'spare' compensates for the risk of destruction of the elongated penis of earwigs (Insecta: Dermaptera). *Naturwissenschaften*, 88(11), 468–71

Scharf, I., and Martin, O. Y. (2013). Same-sex sexual behavior in insects and arachnids: prevalence, causes, and consequences. *Behavioral Ecology and Sociobiology*, 67(11), 1719–30

Sivinski, J. (1978). Intrasexual aggression in the stick insects Diapheromera veliei and D. covilleae and sexual dimorphism in the Phasmatodea. *Psyche: A Journal of Entomology*, 85(4), 395–405

http://www.wired.com/2015/02/50-shades-wrong-disturbing-insect-sex/

http://blogs.discovermagazine.com/notrocketscience/2011/07/15/
hermaphrodite-insects-fertilise-daughters-with-parasitic-sperm/#.
VXlZkhNVhHw

Let's Spend Some Quaility Time Together

Ahlers, C. J., Schaefer, G. A., Mundt, I. A., Roll, S., Englert, H., Willich, S. N., and Beier, K. M. (2011). How unusual are the contents of paraphilias? Paraphilia-associated sexual arousal patterns in a community-based sample of men. *The Journal of Sexual Medicine*, 8(5), 1362–70

Çetinkaya, H., and Domjan, M. (2006). Sexual fetishism in a quail (*Coturnix japonica*) model system: test of reproductive success. *Journal of Comparative Psychology*, 120(4), 427–32

Köksal, F., Domjan, M., Kurt, A., Sertel, Ö., Örüng, S., Bowers, R., and Kumru, G. (2004). An animal model of fetishism. *Behaviour Research and Therapy*, 42(12), 1421–34

http://www.cracked.com/blog/7-sexy-japanese-game-shows-that-will-make-you-hate-sex/

https://en.wikipedia.org/wiki/Eurotrash_(TV_series)

http://www.huffingtonpost.com/2013/10/23/sexual-fetish_n_4144418.html

A Sick Idea

Tanaka, Y., Yoshimura, J., Simon, C., Cooley, J. R., and Tainaka, K. I. (2009). Allee effect in the selection for prime-numbered cycles in periodical cicadas. *Proceedings of the National Academy of Sciences*, 106(22), 8975–9

You've Got To Hide Your Bug Away

Penacchio, O., Lovell, P. G., Cuthill, I. C., Ruxton, G. D., and Harris, J. M. (2015). Three-dimensional camouflage: exploiting photons to conceal form. *The American Naturalist*, 186(4), 553–63

Should I Stay or Should I Go?

Blanchard, T. C., and Hayden, B. Y. (2015). Monkeys are more patient in a foraging task than in a standard intertemporal choice task. *PloS one*, 10(2), e0117057

Cassini, M. H., Kacelnik, A., and Segura, E. T. (1990). The tale of the screaming hairy armadillo, the guinea pig and the marginal value theorem. *Animal Behaviour*, 39(6), 1030–50

Constantino, S. M., and Daw, N. D. (2015). Learning the opportunity cost of time in a patch-foraging task. *Cognitive, Affective, & Behavioral Neuroscience*, 15(4), 837–53

Naef-Daenzer, B. (2000). Patch time allocation and patch sampling by foraging great and blue tits. *Animal Behaviour*, 59(5), 989–9

Great Tits and Seedy Locations: Test

Brodin, A., and Lundborg, K. (2003). Rank-dependent hoarding effort in
willow tits (*Parus montanus*): a test of theoretical predictions. *Behavioral
Ecology and Sociobiology*, 54(6), 587–92

Brodin, A., and Urhan, A. U. (2014). Interspecific observational memory in
a non-caching *Parus* species, the great tit *Parus major*. *Behavioral Ecology
and Sociobiology*, 68(4), 649–56

Brodin, A., and Urhan, A. U. (2015). Sex differences in learning ability in a
common songbird, the great tit—females are better observational learn-
ers than males. *Behavioral Ecology and Sociobiology*, 69(2), 237–41

Lucas, J. R., Brodin, A., de Kort, S. R., and Clayton, N. S. (2004). Does hippo-
campal size correlate with the degree of caching specialisation? *Proceedings of the Royal Society of London B: Biological Sciences*, 271(1556), 2423–9

The Pecking Order

Chiao, J. Y., Mathur, V. A., Harada, T., and Lipke, T. (2009). Neural basis of
preference for human social hierarchy versus egalitarianism. *Annals of the
New York Academy of Sciences*, 1167(1), 174–81

Hogue, M. E., Beaugrand, J. P., and Lagué, P. C. (1996). Coherent use of information by hens observing their former dominant defeating or being
defeated by a stranger. *Behavioural Processes*, 38(3), 241–52

Pratto, F., Sidanius, J., Stallworth, L. M., and Malle, B. F. (1994). Social dominance orientation: a personality variable predicting social and political
attitudes. *Journal of Personality and Social Psychology*, 67(4), 741

Pratto, F., Çidam, A., Stewart, A. L., Zeineddine, F. B., Aranda, M., Aiello,
A., . . . and Henkel, K. E. (2013). Social dominance in context and in individuals' contextual moderation of robust effects of social dominance
orientation in 15 languages and 20 countries. *Social Psychological and Personality Science*, 4(5), 587–99

A Weighty Problem

Lentink, D., Haelsteiner, A. F., and Ingersoll, R. (2015). *In vivo* recording of
aerodynamic force with an aerodynamic force platform: from drones to
birds. *Journal of the Royal Society Interface*, 12(104), 20141283

https://www.newscientist.com/article/dn26807-if-birds-in-a-truck-fly-
does-the-truck-get-lighter/

For Eagle-Eyed Readers: Test

Dennett, D. C. (2013). *Intuition Pumps and Other Tools for Thinking*. New York: W. W. Norton & Co.

Kram, Y. A., Mantey, S., and Corbo, J. C. (2010). Avian cone photoreceptors tile the retina as five independent, self-organizing mosaics. *PLoS ONE*, 5(2), e8992

Are You a Bat, Man?

Buckingham, G., Milne, J. L., Byrne, C. M., and Goodale, M. A. (2015). The size-weight illusion induced through human echolocation. *Psychological Science*, 26(2), 237–42

Kolarik, A. J., Cirstea, S., Pardhan, S., and Moore, B. C. J. (2014). A summary of research investigating echolocation abilities of blind and sighted humans. *Hearing Research*, 310, 60–68. doi:10.1016/j.heares.2014.01.010

Neuweiler, G. (1984). Foraging, echolocation and audition in bats. *Naturwissenschaften*, 71(9), 446–55

Thaler, L., Arnott, S. R., and Goodale, M. A. (2011). Neural correlates of natural human echolocation in early and late blind echolocation experts. *PLoS One*, 6(5), e20162

An Elephant Never Forgets

Dungl, E., Schratter, D., and Huber, L. (2008). Discrimination of face-like patterns in the giant panda (*Ailuropoda melanoleuca*). *Journal of Comparative Psychology*, 122(4), 335–43

Martin, F., and Niemitz, C. (2003). 'Right-trunkers' and 'left-trunkers': side preferences of trunk movements in wild Asian elephants (*Elephas maximus*). *Journal of Comparative Psychology*, 117(4), 371–9

Photos from Flickr users Kevin Dooley (https://www.flickr.com/photos/pagedooley/)

and David Schroeter (https://www.flickr.com/photos/drs2biz/), used under a Creative Commons licence (CC BY 2.0): https://creativecommons.org/licenses/by/2.0/legalcode

A Shrewd Judgement

Hopkins, W. D. (2013). Neuroanatomical asymmetries and handedness in chimpanzees (*Pan troglodytes*): a case for continuity in the evolution of hemispheric specialisation. *Annals of the New York Academy of Sciences*, 1288(1), 17–35

Hopkins, W. D., Schaeffer, J., Russell, J. L., Bogart, S. L., Meguerditchian, A., and Coulon, O. (2015). A comparative assessment of handedness and its potential neuroanatomical correlates in chimpanzees (*Pan troglodytes*) and bonobos (*Pan paniscus*). *Behaviour*, 152(3–4), 461–92

Langergraber, K. E., Prüfer, K., Rowney, C., Boesch, C., Crockford, C., Fawcett, K., . . . and Vigilant, L. (2012). Generation times in wild chimpanzees and gorillas suggest earlier divergence times in great ape and human evolution. *Proceedings of the National Academy of Sciences*, 109(39), 15716–21

Maille, A., Jäschke, N., Joly, M., Scheumann, M., Blois-Heulin, C., and Zimmermann, E. (2013). Does a nonprimate mammal, the Northern tree shrew (*Tupaia belangeri*), exhibit paw preference in two forms of a grasping task? *Journal of Comparative Psychology*, 127(1), 14–23

Left Behind?

Beaton, A. A., Kaack, I. H., and Corr, P. J. (2015). Handedness and behavioural inhibition system/behavioural activation system (BIS/BAS) scores: a replication and extension of Wright, Hardie, and Wilson (2009). *Laterality: Asymmetries of Body, Brain and Cognition* 20(5), 585–603

Braccini, S. N., and Caine, N. G. (2009). Hand preference predicts reactions to novel foods and predators in marmosets (*Callithrix geoffroyi*). *Journal of Comparative Psychology*, 123(1), 18–25

Carver, C. S., and White, T. L. (1994). Behavioral inhibition, behavioral activation, and affective responses to impending reward and punishment: the BIS/BAS scales. *Journal of Personality and Social Psychology*, 67(2), 319

Davidson, R. J. (2004). Well-being and affective style: neural substrates and biobehavioural correlates. *Philosophical Transactions-Royal Society of London Series B Biological Sciences*, 35, 1395–1412

Hardie, S. M., and Wright, L. (2014). Differences between left-and right-handers in approach/avoidance motivation: influence of consistency of handedness measures. *Frontiers in Psychology*, 18(5), 520–35

Cuckoo Clocks His Rivals

Kilgallon, S. J., and Simmons, L. W. (2005). Image content influences men's semen quality. *Biology Letters*, 1(3), 253–5

Pham, M. N., Shackelford, T. K., Holden, C. J., Zeigler-Hill, V., Hummel, A., and Memering, S. L. (2014). Partner attractiveness moderates the relationship between number of sexual rivals and in-pair copulation frequency in humans (*Homo sapiens*). *Journal of Comparative Psychology*, 128(3), 328–31

Pizzari, T., Cornwallis, C. K., Løvlie, H., Jakobsson, S., and Birkhead, T. R. (2003). Sophisticated sperm allocation in male fowl. *Nature*, 426(6962), 70–74

Dad Calls It Quits

Alvergne, A., Faurie, C., and Raymond, M. (2009). Father–offspring resemblance predicts paternal investment in humans. *Animal Behaviour*, 78(1), 61–9

Diniz, P., Ramos, D. M., and Macedo, R. H. (2015). Attractive males are less than adequate dads in a multimodal signalling passerine. *Animal Behaviour*, 102, 109–17

Guéguen, N. (2014). Cues of men's parental investment and attractiveness for women: a field experiment. *Journal of Human Behavior in the Social Environment*, 24(3), 296–300

School of Collective Nouns

http://www.thelinguafile.com/2015/05/a-destruction-of-cats-collective-nouns.html#.VYq91RNVhHw

http://www.collectivenouns.biz/list-of-collective-nouns/collective-nouns-people/

http://editra.ca/DEMZ/text/Noms-collectifs-designant-des-assemblages-d-animaux.html

http://www.oxforddictionaries.com/words/what-do-you-call-a-group-of

Bird Is the Word (or Starlings in Their Eyes)

Akçay, Ç., Anderson, R. C., Nowicki, S., Beecher, M. D., and Searcy, W. A. (2015). Quiet threats: soft song as an aggressive signal in birds. *Animal Behaviour*, 105, 267–74

Brumm, H., and Todt, D. (2004). Male–male vocal interactions and the adjustment of song amplitude in a territorial bird. *Animal Behaviour*, 67(2), 281–6

Doolittle, E. L., Gingras, B., Endres, D. M., and Fitch, W. T. (2014). Overtone-based pitch selection in hermit thrush song: unexpected convergence with scale construction in human music. *Proceedings of the National Academy of Sciences*, 111(46), 16616–21

Gerhardt, H. C., and Huber, F. (2002). *Acoustic Communication in Insects and Anurans: Common Problems and Diverse Solutions*. Chicago, IL: University of Chicago Press

Goodwin, S. E., and Podos, J. (2014). Team of rivals: alliance formation in territorial songbirds is predicted by vocal signal structure. *Biology Letters*, 10(2), 20131083

Mennill, D. J., Ratcliffe, L. M., and Boag, P. T. (2002). Female eavesdropping on male song contests in songbirds. *Science*, 296(5569), 873

Nowicki, S., and Searcy, W. A. (2004). Song function and the evolution of female preferences: why birds sing, why brains matter. *Annals of the New York Academy of Sciences*, 1016(1), 704–23

Stringing You Along

Jacobs, I. F., and Osvath, M. (2015). The string-pulling paradigm in comparative psychology. *Journal of Comparative Psychology*, 129(2), 89–120

Krasheninnikova, A., Bräger, S., and Wanker, R. (2013). Means–end comprehension in four parrot species: explained by social complexity. *Animal Cognition*, 16(5), 755–64

Osthaus, B., Lea, S. E., and Slater, A. M. (2005). Dogs (*Canis lupus familiaris*) fail to show understanding of means-end connections in a string-pulling task. *Animal Cognition*, 8(1), 37–47

Are You Smarter than an Orang-Utan? Test

Swartz, K. B., Himmanen, S. A., and Shumaker, R. W. (2007). Response strategies in list learning by orangutans (*Pongo pygmaeus* × P. abelii). *Journal of Comparative Psychology*, 121(3), 260–69

Are You Smarter than a Chimpanzee? #2

Taubert, J., Qureshi, A. A., and Parr, L. A. (2012). The composite face effect in chimpanzees (*Pan troglodytes*) and rhesus monkeys (*Macaca mulatta*). *Journal of Comparative Psychology*, 126(4), 339

Are You Smarter than a Chimpanzee? #3

Parr, L. A., and Waller, B. M. (2006). Understanding chimpanzee facial expression. Insights into the evolution of communication. *Social Cognitive and Affective Neuroscience*, 1(3), 221–8

Parr, L. A., Waller, B. M., Vick, S. J., and Bard, K. A. (2007). Classifying chimpanzee facial expressions using muscle action. *Emotion*, 7(1), 172

Prüfer, K., Munch, K., Hellmann, I., Akagi, K., Miller, J. R., Walenz, B., . . . and Pääbo, S. (2012). The bonobo genome compared with the chimpanzee and human genomes. *Nature*, 486(7404), 527–31

It's Been Emotional

Bandler, R., Keay, K. A., Floyd, N., and Price, J. (2000). Central circuits mediating patterned autonomic activity during active vs. passive emotional coping. *Brain Research Bulletin*, 53(1), 95–104

Dawkins, M. S. (2000). Animal minds and animal emotions. *American Zoologist*, 40(6), 883–8

Fodor, J. (2007). Why pigs don't have wings. *London Review of Books*, 29(20)

Plutchik, R. (1980). *Emotion: A Psychoevolutionary Synthesis*. New York: Harper Collins

Scherer, K. R., & Wallbott, H. G. (1994). Evidence for universality and cultural variation of differential emotion response patterning. *Journal of Personality and Social Psychology*, 66(2), 310.

Soltysik, S., and Jelen, P., 2005. In rats, sighs correlate with relief. *Physiology and Behavior* 85, 598–602

Psycho Gorilla, Qu'est-ce que c'est?

Latzman, R. D., Drislane, L. E., Hecht, L. K., Brislin, S. J., Patrick, C. J., Lilienfeld, S. O., . . . and Hopkins, W. D. (2015). A chimpanzee (*Pan troglodytes*) model of triarchic psychopathy constructs development and initial validation. *Clinical Psychological Science*, 4(1), 50–66

Every Body Hurts?

Sneddon, L. U., Elwood, R. W., Adamo, S. A., and Leach, M. C. (2014). Defining and assessing animal pain. *Animal Behaviour*, 97, 201–12

Oh, Beehave

Schmidt, J. O., Blum, M. S., and Overal, W. L. (1983). Hemolytic activities of stinging insect venoms. *Archives of Insect Biochemistry and Physiology*, 1(2), 155–60

Smith, M. L. (2014). Honey bee sting pain index by body location. *PeerJ*, 2, e338 http://www.westsoundbees.org/newsletters/june_2008.pdf

Taking the P***

Yang, P. J., Pham, J., Choo, J., and Hu, D. L. (2014). Duration of urination does not change with body size. *Proceedings of the National Academy of Sciences*, 111(33), 11932–7

Self-Awareness: Have Monkeys Cottoned On?

Gallup Jr, G. G. (1998). Self-awareness and the evolution of social intelligence. *Behavioural Processes*, 42(2), 239–47

Gallup Jr, G. G., Anderson, J. R., and Shillito, D. J. (2002). The mirror test. In *The Cognitive Animal: Empirical and Theoretical Perspectives on Animal Cognition*. Cambridge, MA: MIT Press, pp. 325–33

Gallup Jr, G. G., Burch, R. L., and Platek, S. M. (2002). Does semen have antidepressant properties? *Archives of Sexual Behavior*, 31(3), 289–93

Hare, B., Call, J., and Tomasello, M. (2006). Chimpanzees deceive a human competitor by hiding. *Cognition*, 101(3), 495–514

Hauser, M. D., Miller, C. T., Liu, K., and Gupta, R. (2001). Cotton-top tamarins (*Saguinus oedipus*) fail to show mirror-guided self-exploration. *American Journal of Primatology*, 53(3), 131–7

Legrain, L., Cleeremans, A., and Destrebecqz, A. (2011). Distinguishing three levels in explicit self-awareness. *Consciousness and Cognition*, 20(3), 578–85

Plotnik, J. M., De Waal, F. B., and Reiss, D. (2006). Self-recognition in an Asian elephant. *Proceedings of the National Academy of Sciences*, 103(45), 17053–7

Prior, H., Schwarz, A., and Güntürkün, O. (2008). Mirror-induced behavior in the magpie (*Pica pica*): evidence of self-recognition. *PLoS Biology*, 6(8), e202

Reiss, D., and Marino, L. (2001). Mirror self-recognition in the bottlenose dolphin: a case of cognitive convergence. *Proceedings of the National Academy of Sciences*, 98(10), 5937–42

Soler, M., Pérez-Contreras, T., and Peralta-Sánchez, J. M. (2014). Mirror-mark tests performed on jackdaws reveal potential methodological problems in the use of stickers in avian mark-test studies. *PloS one*, 9(1), e86193

http://www.sciencemag.org/news/2011/07/behavioral-scientist-marc-hauser-resigns-harvard?_ga=1.85701401.1051845032.1449232664

http://www.sciencemag.org/news/2014/05/harvard-misconduct-investigation-psychologist-released

http://www.sciencemag.org/news/2012/09/harvard-psychology-researcher-committed-fraud-us-investigation-concludes

Epilogue: Relative Values Revisited

Dennett, D. C. (2013). *Intuition Pumps and Other Tools for Thinking*. New York: W. W. Norton & Co.

Nagel, T. (1974). What is it like to be a bat? *The Philosophical Review*, 83(4), 435–50

http://fcmconference.org/img/CambridgeDeclarationOnConsciousness.pdf

https://aeon.co/essays/how-consciousness-works-and-why-we-believe-in-ghosts

https://www.ted.com/talks/dan_dennett_on_our_consciousness/transcript?language=en

Credits

The Usual Waspects © 2014 José Lino-Net. PLOS ONE version © 2014 de Souza et al. Reproduced under Creative Commons Attribution License from de Souza et al (2014). See References section for full citation. Special thanks are due to André Rodrigues and José Lino-Neto for supplying high resolution versions of the wasp photographs; A Stable Personality Canine-Big-Five Inventory © Sam Gosling. We thank Sam Gosling for kindly supplying a copy of the test, and for granting permission to reproduce it; Dog-Person or Person-Dog © 2014 de Kujala et al. Reproduced under Creative Commons Attribution License from de Souza et al (2014). See References section for full citation. Special thanks are due to Jan and Miiamaaria Kujala for supplying the original high-resolution photographs, to which they retain all rights; Losing Your Marbles © 2016 Ben Ambridge. We thank unanimous.ai for their generous help with running the online guessing game; Circle of Life Ebbinghaus illusion courtesy of Marco Bertamini, University of Liverpool; Something to Crow About? © 2014 APA. Reprinted with permission from Jelbert et al (2015). See References section for full citation; Students vs Squirrels, Sorta © 2009 APA. Reprinted with permission from Wills et al (2009). See References section for full citation; Box Clever © 2013 APA. Reprinted with permission from Leighty et al (2013). See References section for full citation; Oh, What a Tangled Web We Weave © 1971 John Wiley & Sons, Ltd. Reprinted with permission from Witt (1971). See References section for full citation (journal is now *Systems Research and Behavioral Science*); You've Got To Hide Your Bug Away © Olivier Penacchio. Used under Creative Commons Attribution-Share Alike 3.0 Unported license. See https://commons.wikimedia.org/wiki/File:Actias_luna_back_uppermost.JPG and https://commons.wikimedia.org/wiki/File:Actias_luna_upside_down.JPG; The Pecking Order © 1994 APA. Questionnaire adapted with permission from Pratto et al (1994). See References section for full citation; An Elephant Never Forgets © 2006 David Schroeter. Reproduced by kind permission of David Schroeter © 2009 Kevin Dooley. Reproduced under Creative Commons CC BY 2.0 License. See https://creativecommons.org/licenses/by/2.0/legalcode; Left Behind © 1994 APA. Reprinted with permission from Carver and White (1994). See References section for full citation; Cuckoo Clocks his Rivals © 2014 APA. Questionnaire reprinted with permission from Pham et al (2014). See References section for full citation; Dad Calls it Quits © 2014 APA. Questionnaire reprinted with permission from Pham et al (2014). See References section for full citation (under Cuckoo Clocks his Rivals); Are You Smarter Then a Chimpanzee? #3 © 2006, Oxford University Press. Reprinted with permission from Parr and Waller (2006). See References section for full citation; It's Been Emotional © 1994 APA. Table adapted from Scherer and Wallbott (1994). See References section for full citation; Psycho Gorilla, Qu'est-ce Que C'est? © 2016 Association for Psychological Science / Sage. Questionnaire reproduced with permission from Latzman et al (2016). See References section for full citation.